Frank Barnaby is a nuclear physicist by training. He worked at the Atomic Weapons Research Establishment, Aldermaston and University College, London. He was Director of the Stockholm International Peace Research Institute (SIPRI) from 1971 to 1981 and Guest Professor at the Free University, Amsterdam. He currently works for the Oxford Research Group on research into military technology, civil and military nuclear issues, and the terrorist use of weapons of mass destruction.

How to Build a Nuclear Bomb

and other weapons of mass destruction

FRANK BARNABY

NATION BOOKS
New York

For Wendy, Sophie, and Ben

HOW TO BUILD A NUCLEAR BOMB *and Other Weapons of Mass Destruction*

Copyright © 2004 by Frank Barnaby

Published by
Nation Books
An Imprint of Avalon Publishing Group Incorporated
245 West 17th St., 11th Floor
New York, NY 10011

Nation Books is a co-publishing venture of the Nation Institute
and Avalon Publishing Group Incorporated.

Library of Congress Cataloging-in-Publication Data is available.

ISBN 1-56025-603-6

9 8 7 6 5 4 3 2 1

Printed in the United States of America
Distributed by Publishers Group West

Contents

PREFACE to the US edition (2004)

Former Iraqi President Saddam Hussein is out of action, having been captured on December 14, 2003, by American forces at Ad-Dawr, about eleven miles from his hometown of Tikrit. Yet weapons of mass destruction (WMDs) remain a, if not *the*, main cause of concern in the international community, especially among leading Western powers, fearful of nuclear proliferation in the Middle East, a fear enhanced by evidence of nuclear-weapons programs in North Korea, Iran, and Libya, and anxious about the nuclear standoff between India and Pakistan. However that concern has been complicated in the aftermath of the Anglo-American invasion of Iraq.

When they went to war against Iraq on March 20, 2003, both US President George W. Bush and British Prime Minister Tony Blair invoked the scenario of Saddam Hussein deploying chemical and biological weapons. Saddam, an irresponsible and unpredictable leader, could not, they argued, be trusted with WMDs. Disarming Iraq of its WMDs and ballistic missiles was given as the primary reason for going to war. Yet as of writing no WMDs have been found, while a political firestorm has erupted in Britain about the alleged "sexing up" of intelligence reports on Iraqi's WMDs to sell the war to a skeptical British and American public. One casualty of the scandal was the suicide of Dr. David Christopher Kelly.

The Kelly tragedy made much more acute the deep division in British society about participating in the Iraq war. Fifty-nine-year-old David Kelly was a leading international expert in biological weapons and warfare. As a United Nations weapons inspector in Iraq, between 1991 and 1998, he made thirty-seven visits to the country, investigating Iraq's former biological-weapons program. He also led all the inspections at Russia's biological-warfare facilities from 1991 to 1994 under the 1992 US, UK, and Russian agreement.

A microbiologist educated at the universities of Leeds, Birmingham, and Oxford he worked at the British Ministry of Defence's chemical-weapons research center at Porton Down, Wiltshire, becoming the head of microbiology before joining the Ministry of Defence and the Foreign Office as a consultant on arms control. Part of his job was to brief journalists on defense issues.

A quiet man who normally shunned the limelight, Dr. Kelly was thrust into the media spotlight in July 2003 after he was identified in the press as the man the Blair government believed leaked information about Iraq's WMD program to BBC reporter Andrew Gillingham. Kelly soon became embroiled in the furious row between the government and the BBC over claims that the British government's dossier on Iraq's illegal WMD capabilities, published on September 24, 2002 to mobilise public support for the coming invasion of Iraq, was "sexed up."

On July 15, 2003, Kelly was called to give evidence at the House of Commons Foreign Affairs Select Committee. He told the Committee that he was not the source of the "sexed up" claim. On July 17, Kelly's family contacted the police after he failed to return to his home in Oxfordshire. On July 18, his body was found in the countryside a few miles from his home. He had bled to death from a wound in his left wrist, in an apparent suicide. On July 20, the BBC confirmed that Kelly was indeed the source of the Gillingham's report about the "sexed up" dossier.

Dr. Kelly's death caused considerable public disquiet and criticism of Tony Blair and his government. The dignified way his wife Janice and three daughters handled the tragedy increased public concern. Kelly was clearly upset about his treatment and the hostile interrogation at the parliamentary committee, and worried that the affair may have consequences for his pension. But many believed that David Kelly was made a scapegoat to divert public attention from the government's role in the affair. The circumstances of his death and the fact that David Kelly had, for a number of years, been a committed member of the Baha'i faith that condemns suicide added to the disquiet. Tony Blair bowed to public pressure and set up an inquiry, headed by Lord Hutton, to investigate the circumstances of Dr. Kelly's death. The events surrounding his death and the public sentiment against the Iraq war have severely dented Tony Blair's popularity. Only time will tell whether his premiership can survive.

In the United States, ambassador Joseph Wilson's accusation that the Bush Administration manipulated intelligence about Saddam Hussein's weapons programs to justify an invasion of Iraq, along with the growth of American casualties there, has produced a certain cynicism about the war that even the capture of Saddam has only partially allayed.

Bush and Blair's problems have been compounded by the fact that despite the great efforts made since June 2003 to find them, WMDs have yet to be found in Iraq. David Kay, the head of the 1,200-strong Iraq Survey Group, the coalition's team dispatched to find WMDs in Iraq, has said that he plans to leave before the ISG's work is completed. The ISG's interim report, published in October, said that the team was unable to find WMDs or any active program to develop or produce them.

In May 1991, after the first Gulf war, the United Nations Special Commission and the International Atomic Energy Agency—the two international organizations responsible for finding and destroying Iraq's chemical-, biological-, and nuclear-weapons programs, its WMD arsenals and its stocks of chemical and biological agents—began their work. In December 1998, the UN withdrew its inspectors because Iraq failed to cooperate fully with them, thus paving the way for President Clinton to bomb Iraq.

It now seems reasonable to assume that the inspectors succeeded in their task, at least as far as finding and destroying Iraq's militarily significant chemical and biological munitions is concerned, and that Iraq did not fabricate more WMDs after 1998. Whether the Iraqis were able to squirrel away significant stocks of chemical or biological agents is not clear.

Saddam Hussein undoubtedly spent large sums on WMD programs until the 1991 Gulf war but after his defeat his main objective may well have been survival in power. He may have decided that deploying WMDs threatened rather than helped his survival. If this is true, the Iraqis had no WMDs when the 2003 Iraq war began. They may have had small stocks of biological warfare agents but no munitions.

However, the ISG has found evidence that Iraq intended to maintain its capability to develop WMDs in the future, including the preservation of biological research capabilities and strains of bacteria to be used in the future production of biological weapons. There was also evidence of contacts with North Korea about possible future development of long-range ballistic missiles. In other words, there was evidence of Iraqi intention rather than Iraqi capability.

In an interview on December 16, 2003, President Bush was asked why he had stated that Saddam Hussein had such weapons when it appears that they only had the intention to acquire them. He replied: "So what's the difference? If he were to acquire weapons, he would be the danger." Many believe that there is a difference; a country that has actually deployed an effective WMD force is clearly a greater threat to potential adversaries than one that just has the intention to develop WMDs at some future date.

In fact Bush's rather blasé attitude may actually hinder international cooperation and diplomatic attempts to control and eventually seek to abolish WMDs. What is more, the Anglo-American exaggeration about Saddam's WMD potentiality could also cloud the real issues about the danger of WMDs. Ironically, Saddam could turn out to be the least of our problems.

During a visit to Iraq's nuclear establishment at Tuwaitha in 1979 I met a number of the nuclear scientists working there. Some were of very high quality, educated at top universities in the United States, England and other countries, and at the international nuclear center at CERN, in Geneva, Switzerland. They were obviously capable of designing nuclear weapons.

It is a sobering thought that a country needs only a very small group of top nuclear scientists to design and develop nuclear weapons. The United Nations inspectors discovered that Iraq had developed a design for an effective nuclear weapon and had put together and tested its nonnuclear components. In fact, in a rare moment of self-reflection, Saddam Hussein has regretted that he didn't develop a nuclear missile system before he invaded Kuwait in 1991. He must have been thinking of the North Korean example, where a crisis over WMDs was handled very differently.

In a speech on December 16, 2003, at the Monterey Institute of International Studies, William Perry, US Defense Secretary in the Clinton Administration, warned that North Korea was a greater nuclear threat than Iraq. He also stated that in the next decade, a nuclear terrorist device could be exploded in an American city. This is likely unless the United States "develops more effective safeguards against the spread of the fearsome weapons."

"No one would doubt that Al-Qaeda would execute that nightmare scenario if they could get their hands on nuclear weapons," he said. Preventing terrorists from acquiring nuclear weapons should be the "acid test" of America's security policy. It is a test, Perry said, that the country is failing.

Diplomacy, in his opinion, is the only way to prevent North Korea from developing nuclear weapons. As Defense Secretary, Perry handled the 1994 nuclear crisis that brought the Korean peninsula to the brink of war. Although as it was reported, "he had a plan on his desk to bomb North Korea's Yongbyon nuclear facility, the Americans negotiated a regional agreement for North

Korea to cease its nuclear weapons program in exchange for two nuclear reactors that could not be used to develop nuclear weapons." Unfortunately, the agreement was never fully implemented and North Korea resumed its nuclear-weapons program after President Clinton left office.

Another country with ambitions to acquire nuclear weapons is Iran. It is generally believed that Iran has two previously secret nuclear facilities that may be part of a nuclear-weapons program. The Iranian government has acknowledged the existence of the facilities but claims they are part of its civil nuclear program and that it does not have a military nuclear programme.

It is well-known that Iran has a civilian nuclear-power reactor under construction. The 1,000-megawatt light-water reactor is being built at Bushehr by the Russians. It will use low-grade enriched uranium as fuel. Under the contract Iran has with Russia, Russia will provide the fuel for the lifetime of the reactor and will take the spent fuel back to Russia for storage and possibly reprocessing. This power reactor is, according to Iran, the first of a series of power reactors planned to generate 6,000 megawatts of electricity.

Iran operates four research reactors, three at the Estahan Nuclear Technology Center and one at the Nuclear Research Center in Teheran. Two, at Estahan, are subcritical assemblies used for training nuclear physicists and technicians; they have both been operating since 1992. The third at Estahan is a 30-kilowatt-thermal research reactor used for research purposes; it has been operating since 1994. The fourth is a 5-megawatt thermal reactor also used for research; it has been operating since 1967, an indication of the length of time during which Iran has been interested in nuclear technology.

The two facilities suspected of being part of a nuclear-weapons program are a plant to produce heavy water, located near the town of Arak, about 250 kilometers from Teheran; and a gas centrifuge plant for enriching uranium, under construction at Natanz, forty kilometers from Kashan. Very few details of this plant are publicly known.

Heavy water (water in which the hydrogen is the deuterium

isotope) is a very good moderator and coolant for a reactor fueled with natural uranium. Such a reactor is excellent for the production of plutonium of a grade suitable for use in very effective nuclear weapons (so-called weapons-grade plutonium). The Dimona reactor used by Israel to produce plutonium for its nuclear weapons is a heavy water–natural uranium reactor, as is the Cirus reactor used by India produce plutonium for its nuclear weapons.

Heavy water and enriched uranium can be used both in civil and military nuclear programs; they are dual-use materials. For example, the Candu-type civil nuclear-power reactor developed and used by Canada uses heavy water and a gas-centrifuge plant, which can produce the low-enriched uranium needed to fuel civil nuclear-power reactors.

The production of heavy water on a reasonable scale is a much easier task than using a gas centrifuge to produce significant amounts of highly-enriched uranium of the type needed for nuclear weapons. An Iranian facility containing, say, 3,000 gas centrifuges could produce about forty kilograms of highly enriched uranium per year. It would take this facility at least five years to produce enough highly enriched uranium for a nuclear force of six nuclear weapons. For comparison, it is believed that Israel has between 200 and 400 nuclear weapons.

Assuming that about 60 percent of the centrifuges have to be rejected as substandard (a reasonable assumption), Iran would need to produce about 5,000 centrifuges for the facility. Moreover, gas centrifuges break down frequently because of the mechanical stresses they endure. A steady supply of replacement machines must therefore be produced.

A facility operating a cascade of 3,000 centrifuges would use as much electricity as a largish city. It would, therefore, be impossible to operate such a facility clandestinely. Building and effectively operating a gas-centrifuge facility of a useful size is not a trivial task—it is an industrial undertaking. It would probably take Iran at least four or five years to build such a facility and begin producing significant amounts of highly enriched uranium.

Iraq, North Korea, and Iran have shown that if it takes political will to do so, a developing country is able indigenously to construct and operate the complex and sophisticated facilities required to produce the fissile materials (highly enriched uranium and plutonium), needed to fabricate nuclear weapons. They can, if they decide to do so, also produce the agents needed to produce chemical and biological weapons, and to design and produce ballistic missiles to deliver WMDs.

Preventing terrorist groups is considerably more difficult than preventing their spread to countries that do not now have them. It is hard to be optimistic that democracies can succeed in preventing terrorist groups from attacking them with WMDs, including nuclear weapons. History shows that effective counterterrorism is an exceedingly difficult activity.

The ability of the intelligence community to identify and predict threats of terrorist attacks is crucial if such attacks are to be prevented. Monitoring the communications of terrorist groups—the activity known as signal intelligence (SIGINT)—has in the past been used effectively in counterterrorism activities. But today's terrorists can protect their communication systems by the use of, for example, encryption. Human intelligence—HUMINT—is, therefore, the mainstay of counter-terrorism. Experience shows, however, that infiltrating fundamental terrorist groups is, to say the least, extremely difficult.

Rivalries between intelligence agencies within countries and lack of cooperation in intelligence matters among countries seriously reduce the effectiveness of intelligence. One person with adequate access to the political leadership should lead intelligence agencies within countries. International cooperation among national intelligence agencies is essential, as is the integration of national data banks. And so is an effective flow of information to regional and international authorities. International cooperation and flexibility are the keys to good counterterrorism intelligence.

The monitoring and control of the trade, within and among states, in the materials needed by terrorists to fabricate chemical, biological, and nuclear weapons is crucial, and should be considerably improved. Some materials, such as plutonium, should simply not be used and activities, like the reprocessing of spent nuclear-power-reactor fuel, should be stopped.

Perhaps the best we can do is implement as best we can counterterrorist measures and, at the same time, put into place the most effective emergency services we can afford in order to cope with a terrorist attack if it occurs. The post-9/11 establishment in the United States of the Department of Homeland Security is a welcome step in the right direction.

Preface

The twentieth century saw an unprecedented increase in destruction caused by warfare, mainly brought about by the ever-increasing lethality of weapons. The terrorist attacks on New York and Washington on September 11, 2001, and the responses to them, suggest that in the twenty-first century we shall continue to witness large-scale violence by both states and sub-state groups. The destructiveness of warfare and of international terrorism is likely to increase dramatically in the coming years, mainly because of the spread of weapons of mass destruction (WMDs) to countries and international terrorist groups which do not currently have them.

One of the ways the world changed forever on September 11, was that a shocked public realized that international terrorists are prepared to attack even the most powerful and heavily armed country in the world, killing large numbers of people in suicide attacks. And if this was possible, then new attacks may come at any time and anywhere. It is hardly surprising that the nightmare of international terrorists using WMDs has become so disturbing.

new attacks may come at any time and anywhere

When societies are vulnerable, it is essential that there be informed public debate about the risks and the measures needed to address them. Currently, the debate is far from informed, mainly because of the large amount of inaccurate information and misinformation in circulation. These have created public fear, which in turn has been exacerbated by Western governments' badly thought out and hastily implemented counterterrorist policies. These not only reduce the effectiveness of counterterrorism but actually play into the terrorists' hands, weakening democracy by instituting unnecessary repressive measures such as some of those brought in by the 2001 British Anti-terrorism, Crime and Security Bill.

The purpose of this book, then, is to contribute to informed debate by providing factual information on the characteristics of WMDs—biological, chemical, nuclear, and radiological—and the effects of their use. In Part 1 of the book, I describe the current global arsenals of WMDs, who has these, what they have, and the munitions used to deliver the weapons to their targets. The personnel, facilities and materials needed by a state to fabricate WMDs are explained and I analyze the roles of politicians, scientists, industry, the defence bureaucracy, and the military in WMD programs. In Chapter 5, I present two case studies which discuss Iraq's and North Korea's likely involvement with WMDs. Chapter 6 looks at the international impact following the discovery that a state has a WMD program.

Chapter 7 deals in detail with the potential terrorist use of WMDs, and the following chapter attempts to identify the terrorist groups capable of making and using them. The means and likely success of counterterrorism form the subject of Chapter 9; and finally in Chapter 10, I offer some thoughts on what the future might hold.

Introduction—the state we're in

Weapons of mass destruction (WMDs) have been one of the most prominent topics in the news since the terrorist attacks on New York and Washington on September 11, 2001. Not a day passes without a great deal being said about them in the media, by politicians and other commentators. Much of what politicians say is misinformation, often put about for propaganda purposes, and many reporters misunderstand the issues.

The world's leaders continually warn us of the dangers of WMDs. The Secretary General of the United Nations, Kofi Annan, called the terrorist use of nuclear, biological, and chemical weapons "the gravest threat the world faces." And Tony Blair and George W. Bush have frequently told us that international terrorists and the states that support them—particularly Iran, Iraq, and North Korea—are today's greatest threats to national and global security. They claim that "rogue states" are likely to make WMDs available to terrorists who will act as proxies, using the weapons to attack the states' enemies. War is necessary and justified to remove these threats. Unless the regimes in the accused countries are changed, WMDs may be used with devastating effects. Should we believe these prophecies of doom or are they exaggerated nightmares?

It is impossible to judge the threat of WMDs unless we know

the answers to some key questions which this book will attempt to address. How do biological, chemical, and nuclear weapons differ? What are the effects of the use of these weapons? Which terrorist groups are capable of making and using these weapons? What facilities and capabilities do countries need to fabricate and deliver them? Can democracies deal with the threat of biological, chemical, and nuclear terrorism? What are the most effective counterterrorism measures? What does the future hold in the way of terrorism with WMDs? Is cyberspace under threat of terrorist attack? Which new countries will develop and deploy WMDs?

When the Cold War ended in 1991 there were high hopes that the importance given by political and military leaders to weapons of mass destruction, particularly nuclear weapons, would dramatically decrease. There would then be rapid progress in disarmament leading to the total abolition of these weapons, or so it was believed. But this was not to be: more countries now deploy WMDs than ever before.

As East–West relations deteriorated after the Second World War, concern about a global nuclear war, which could have destroyed the Northern Hemisphere, was widespread. Few people now worry about a global nuclear holocaust, but the possibility of a regional nuclear war remains a real one. The hope that future generations will be saved from the scourge of nuclear conflict has yet to be realized; the vision of mushroom clouds rising over our heads has not gone away.

Nuclear weapons are here to stay

Unfortunately, far-reaching nuclear disarmament has not yet been negotiated and there is no reason to believe that it will be in the foreseeable future. On the contrary, nuclear weapons are now back on the agenda more firmly than at any time since the height of the Cold War. For example, the US National Strategy to Combat Weapons of Mass Destruction, completed at the end of 2002, describes a role for nuclear weapons well into the future, not as part of a nuclear deterrent policy but as part of America's war-fighting strategy. Apparently, the Pentagon has prepared contingency plans to use nuclear weapons pre-emptively against targets in seven or more countries—including China, Iran, Iraq, Libya, Russia, and Syria.

According to the National Strategy:

the Pentagon has prepared contingency plans to use nuclear weapons pre-emptively against targets in seven or more countries

> Weapons of mass destruction—nuclear, biological, and chemical —in the possession of hostile states and terrorists represent one of the greatest security challenges facing the United States. We must pursue a comprehensive strategy to counter this threat in all of its dimensions.
>
> An effective strategy for countering WMD, including their use and further proliferation, is an integral component of the National Security Strategy of the United States of America. As with the war on terrorism, our strategy for homeland security, and our new concept of deterrence, the US approach to combat WMD represents a fundamental change from the past. To succeed, we must take full advantage of today's opportunities, including the application of new technologies, increased emphasis on intelligence collection and analysis, the strengthening of alliance relationships, and the establishment of new partnerships with former adversaries.

In March 2002, the British Minister of Defence announced, for the first time ever, that British nuclear weapons could be used in a first strike and against countries that used biological or chemical weapons against British forces or targets in the UK. Both the American and British governments have now reneged on their security assurance guarantees not to use nuclear weapons against countries that do not have them and which are not allied to a nuclear-weapon power. The constraints on the use of nuclear weapons are weakening as nuclear deterrence gives way to pre-emption.

These new policies have been adopted in spite of the "unequivocal undertaking to accomplish the total elimination" of their nuclear weapons entered into by the USA and the UK along with the other established nuclear-weapon states (China, France and Russia at the 2000 Review Conference of the Non-Proliferation Treaty (NPT). One hundred and eighty-seven countries have ratified the NPT, making it the world's most comprehensive multilateral nuclear arms control treaty.

While the established nuclear-weapon powers claim to be opposed to the spread of weapons of mass destruction, particularly nuclear ones, to other countries, they will not get rid of their own nuclear weapons and reneging on such a universal treaty is likely only to encourage other countries to acquire their own WMDs. American political leaders are even prepared to go to war to prevent such proliferation or to disarm some countries that have acquired WMDs. However, the policy is confused, confusing, and hypocritical: there is no suggestion that action will be taken against India, Israel, Pakistan and North Korea, which have nuclear weapons, or Iran, which is suspected of developing them.

The established nuclear-weapon powers are continually improving the quality of their nuclear weapons and developing technologies to support them. By this behavior they show that they believe that their nuclear weapons have considerable political and military value. How then can they be surprised when other countries want these weapons themselves?

Why countries "go nuclear"

There are a number of reasons which might prompt a state to acquire nuclear weapons. Some countries want them to solve real or perceived security needs. Israel, for example, feared, with some reason, that various Arab countries wanted to destroy it when the country was born in 1948 and for a little time afterwards. Israel was, therefore, intent on developing nuclear weapons, as a deterrent or as a weapon of last resort, and began to do so in the 1950s, finally deploying some in the 1973 war.

Prestige is another reason. The fact that all permanent members of the United Nations Security Council are nuclear-weapon powers is not lost on non-nuclear states. Nuclear weapons can give a state a dominant position in its region. Conversely, the risk of loss of prestige is a reason why countries with nuclear weapons, such as France and the United Kingdom, will not give them up. Political leaders may also want nuclear weapons for internal political reasons—to boost their domestic power or to distract their people from social or economic problems. India may have acquired nuclear weapons partly for this last reason, partly to impress Pakistan, and partly to improve its security against China.

> *Nuclear weapons can give a state a dominant position in its region*

There may also be a "domino effect" in some regions; if one country acquires nuclear weapons, neighboring countries may feel obliged to follow suit. Pakistan, for example, felt itself to be under great pressure to get nuclear weapons when India did so.

Weapons of mass destruction: the next terrorist threat?

Even before Hiroshima and Nagasaki were destroyed by atomic bombs, prophetic observers foresaw that WMDs might one day fall into the hands of terrorist groups as well as states. In April 1945, for example, US Secretary of War Henry Stimson warned Harry Truman that: "the future may see a time when such a weapon may be constructed in secret and used suddenly and effectively with devastating power by a wilful nation or group against an unsuspecting nation or group of much greater size and material power." The weapon to which Stimson referred was one that could destroy a whole city—a WMD.

Stimson could not have foreseen the rise of fundamentalist

terrorism and the ways in which it would threaten advanced societies. Many of the most crucial assets of industrialized societies—like large power stations, fuel dumps, liquid gas storage sites, computer networks, major telecommunication centers, major transport centers—without which the society cannot operate effectively, are highly centralized and, therefore, particularly vulnerable to attack or sabotage by terrorist groups.

All the signs are that during the next decade or two fundamentalist terrorism will increase in frequency and the terrorists will be prepared to use WMDs in their attacks. The risk of attacks on crucial targets will increase; attacks on computer networks (cyberterrorism) and attacks on large nuclear-power stations are particularly likely, and alarming, prospects in the future.

International terrorism and democracy

People, particularly those who live in major cities, fear international terrorism. This fear is fed by the intense media interest in the subject, which reached a peak after the terrorist attacks on September 11, 2001. The first attack from outside in modern times on America's homeland deeply shocked the world and showed with dramatic clarity that the most powerful and most heavily armed country in the world was vulnerable.

The Americans responded vigorously by declaring a wide-ranging "war on terror." The "war" has mainly been fought by American forces but involves the military forces of a number of allies and the active support of security and intelligence agencies from a larger number of other countries. So far, major military action, using air power and special forces, has been the destruction of the Taliban regime and attacks carried out on Al Qaeda in Afghanistan. Another is the war on Iraq, ostensibly to destroy any biological and chemical weapons that Iraq may have retained after the United Nations inspectors left in 1998; another stated aim was "regime change" to topple Saddam Hussein.

Lower profile activities in the war on terrorism include, in the words of Paul Rogers and Scilla Elworthy, "support for anti-insurgency and counterterrorism operations in a number of countries, especially the Philippines, the development of significant US bases in a number of Central Asian countries, and continuing support for the Sharon government in Israel in its actions against Palestinian militants and the Palestinian population of the occupied territories."

The feeling that there is little that can be done to counter international terrorism enhances fear of it. This feeling of helplessness is particularly strong in democracies, which are certainly more vulnerable to terrorism than are authoritarian regimes. (Research by William Lee Eubank and Leonard Weinberg has shown that the likelihood of finding a terrorist group in a democracy is 3.5 times greater than the likelihood of finding one in a country with an authoritarian regime.) A typical authoritarian regime is prepared to use any means, however brutal, to eliminate terrorists, irrespective of whether these ultra-repressive actions are as illegal as those used by the terrorists themselves.

> *the likelihood of finding a terrorist group in a democracy is 3.5 times greater than the likelihood of finding one in a country with an authoritarian regime*

Peter Chalk, an expert in responses to terrorism, puts the case:

> Any liberal response to terrorism has to rest on one over-riding maxim: a commitment to uphold and maintain the rule of law. It is quite obvious that the threat of terrorism can be minimized, if not entirely eliminated, by any state that is prepared to use to their fullest extent the entire range of coercive powers at its disposal. However, to do so would merely be to transplant insurgent terrorism from below with institutionalized and bureaucratized terror from above, destroying in the process any moral or legal claim to legitimacy that the state may have.

The aim of a typical terrorist group is to disrupt and destabilize society, as Paul Wilkinson explains: "Political terrorism may be briefly defined as coercive intimidation. It is the systematic use of murder and destruction, and the threat of murder and destruction in order to terrorize individuals, groups, communities or governments." A democratic government has a basic duty to ensure that its citizens can go about their legitimate business with the minimum of hindrance, which is only possible in the absence of coercion and violence. The terrorist's purpose is, by the use of coercive violence, to prevent the citizen from going about his or her legitimate business. The politics of the terrorist are absolute; those of a democracy are based on compromise. The two are inevitably in conflict.

Reacting or overreacting to terrorism?

The main problem for a democracy faced with terrorism is that it must act against terrorists in ways that are legal and constitutional. A democracy must, therefore, evolve counterterrorism measures that are both effective and publicly acceptable. Peter Chalk has described two characterizations of counterterrorism:

> First there is the criminal-justice model which views terrorism as a crime where the onus of response is placed squarely within the bounds of the state's criminal legal system. Second there is the war model which views terrorism as an act of revolutionary/ guerrilla warfare and where the onus of response is placed on the military and the use of, for instance, special forces, retaliatory strikes, campaigns of retribution and troop deployment. The typical approach adopted by liberal democracies in Europe and North America is to treat terrorism as a crime where prosecution and punishment take place within the rule of law. In other words, the response conforms to the criminal-justice model.

To adopt the "war model" would be to acknowledge the political role of the terrorist and legitimate his actions. This the democracies have, up to now, been generally unwilling to do. Consequently, the military have been brought in only as a last resort, to be deployed in an emergency, such as the use of the British Army in Northern Ireland. The new "war" against international terrorism adopts the war model and is a significant move away from the previous policy.

If a democracy overreacts to terrorism by making significant departures from normal legal and law-enforcement procedures the response will be neither effective nor acceptable to the public. Illegality by the state is seen to match the illegality of the terrorist act, and thereby plays into the hands of the terrorist. Departures from the due process of law (such as the failure to obtain search warrants, extracting confessions by torture, internment without trial, the denial of timely access to lawyers, and illegal detention) are deemed excessively repressive and unacceptable in a democracy and, therefore, likely to prove ultimately ineffective because they undermine the democracy they purport to defend.

There are good reasons for believing that the current American administration is overreacting to terrorism. President Bush, elected by an arguable majority in Florida, a minority of the electorate, and a partisan majority in the Supreme Court, has used his popularity after September 11, 2001 to suppress both criticism and civil liberties. The

the end result of too much repression is a police state; the end result of the appeasement of terrorists is anarchy

treatment of prisoners at the US internment camp in Guantanamo Bay, Cuba, and the refusal to bring them to trial, tell them what they are charged with, allow them access to lawyers, or release them are obvious threats to the strength of American democracy.

The refusal by the state to take decisive legal action against terrorists is equally ineffective. Appeasement is likely to encourage terrorists to further violence and enhance public feelings of insecurity. Paul Wilkinson warns: "If a democratic government caves in to extremist movements and allows them to subvert and openly defy the laws and to set themselves up as virtual rival governments within the state, the liberal democracy will dissolve into an anarchy of competing factions and enclaves." In other words, the end result of too much repression is a police state; the end result of the appeasement of terrorists is anarchy.

There is always a tendency for governments, even in liberal democracies, to adopt extreme measures when terrorists resort to great violence. Even if these measures are permitted constitutionally, they will not be publicly acceptable and effective if they appear to go beyond reasonable limits. For example, in 1971, the British government introduced internment without trial in Northern Ireland. It was severely criticized both domestically and internationally, even though terrorist activity in Northern Ireland had reached unprecedented levels.

If democracies are going to deal effectively with the terrorist threat in ways that do not threaten the democratic way of life of their citizens, all of the counterterrorist measures they adopt must be firmly under the control of civil authorities that are accountable to the people through Parliament. In the words of Peter Chalk:

> The invocation, use and continuance of all counterterrorist measures need to be made subject to constant parliamentary supervision and independent judicial oversight. In order to strike a balanced response that does not unduly restrict or abuse individual rights and freedoms, it is absolutely essential that the state is held accountable for its actions and that mechanisms exist for the redress of grievances. Antiterrorist measures should therefore be formulated according to clear and precise rules so that all concerned are able to make an adequate assessment of their own powers, obligations and duties.

Fear of biological and chemical terrorism

A biological or chemical attack is probably more likely than a nuclear one and public concern about the former type of terrorist attacks has been high since 11 September. (Though if it were widely believed that a nuclear attack was likely, it would probably be more feared than any other type of attack.)

In turn, biological weapons are more feared than chemical ones because populations feel themselves to be particularly vulnerable to them; it is extremely difficult to protect populations, rather than military forces, against biological attack. In addition, we have an atavistic fear of disease, perhaps dating back to past epidemics, such as the Black Death, which between 1346 and 1350 killed one-third of Europe's population, significantly reducing the inhabitants of 200,000 towns and villages. The psychological impact of the chaos and despair that swept the land may be deep in our psyche. For this reason, and given their exposure and vulnerability, it is likely that populations would panic if involved in a biological attack.

A terrorist group, or even an individual with relatively small financial and personnel resources, could construct an effective biological weapon and release it. An advantage of biological weapons for terrorists is that only a small amount of biological agent is required, because

Public fear is increased by well-publicized statements

microorganisms can be reproduced relatively easily. Chemical weapons, including highly lethal nerve agents, are also relatively easy to prepare from readily available chemicals and chemical apparatus.

Public fear is increased by well-publicized statements, particularly by American and British political leaders, linking Iraq, Iran, and North Korea, the "rogue states" forming Bush's "axis of evil," with international terrorist groups. These three countries are portrayed as possessing biological and chemical weapons and likely to make some available to terrorists.

Fear of becoming involved in a biological or chemical attack is pervasive. The publication of plans to protect populations from biological and chemical attacks—by, for example, stockpiling vaccines or vaccinating whole populations or groups at particular risk—does not reassure but simply adds to people's anxiety.

An awesome future possibility is the use of genetic engineering by military scientists and terrorists to produce new and more deadly biological warfare agents. This even raises the prospect of biological warfare against a specific ethnic group using "genetic-homing" weapons that could target, for example, a genetic structure shared by particular ethnic groups.

The best way to overcome fear of WMDs is to grapple with the nature of the threat. The first step is to understand the characteristics of the weapons and the effects of their use.

PART I

Weapons of mass destruction: What they are and what they do

Weapons of mass destruction take biological, chemical, nuclear or radiological form. The use of the term is recent; it is also controversial. Dan Plesch, of London's Royal United Services Institute, for example, points out that NATO still uses the "nuclear, biological, chemical" description, as each type of weapon has very different effects. The creation of the blanket acronym WMD blurs these distinctions; I will, however, use the acronym in this book for convenience, while recognizing the differences.

Crudely put, biological, chemical and nuclear weapons are designed to kill and injure a large number of people. Nuclear weapons have the additional purpose of destroying much of the enemy's property—particularly his cities and industry—or his own strategic nuclear forces. Radiological weapons are intended to contaminate with radioactivity an area of a city, which will then have to be evacuated and decontaminated—a highly disruptive and expensive procedure. (Radiological weapons are, therefore, strictly speaking "weapons of mass disruption" rather than weapons of mass destruction.)

The awesome lethality of a single WMD puts them into a special category. Political leaders believe that some WMDs are so destructive that they are deterrent weapons, preventing an enemy from making a surprise, preemptive attack. That lethality also makes some of them attractive to fundamentalist religious and political terrorist groups, who want to kill as many people as possible in terrorist attacks in order to seize the headlines. A nuclear explosion, in particular, would fit in with apocalyptic visions of Armageddon.

1 Nuclear weapons

Nuclear history

Soon after nuclear fission was discovered by German physicist Otto Hahn in 1938, it was realized that the energy from fission could be used to produce a nuclear explosion. The fear that Germany and/or Japan might succeed in developing nuclear weapons stimulated the Americans to make a massive effort, known as the Manhattan Project, to develop them first. The effort led to the first nuclear explosion—a test carried out in the New Mexico desert in 1945.

Nuclear weapons have been used only twice in anger: Hiroshima was destroyed by a nuclear weapon on August 6, 1945 and Nagasaki was destroyed three days later. Together the two explosions killed a total of about 250,000 people. Many other nuclear weapons have been exploded in tests and to help designers develop new types, from the first nuclear test on July 16, 1945 in the desert near Alamogordo, New Mexico, to the most recent conducted in Pakistan on May 28, 1998.

Seven nuclear-weapon powers—China, France, India, Pakistan, Russia/the Soviet Union, the United Kingdom and the United States—have tested nuclear weapons. These countries are known to have carried out a total of at least 2,052 nuclear tests. Israel, the eighth known nuclear-weapon power, has not, so far as is publicly known, tested a nuclear weapon.

How a nuclear bomb works

A nuclear weapon produces a powerful explosion by releasing a very large amount of energy in a very short time. It works on the same principle as nuclear reactors which produce electricity; in each case, atoms of uranium or plutonium are split (undergo fission) in a chain reaction. The fission chain reaction in a nuclear reactor is controlled; in a nuclear weapon it is not.

Nuclear fission occurs in different forms of a heavy element—in practice, uranium or plutonium—when a neutron enters the nucleus of an atom of one of these isotopes. When fission occurs the original nucleus is split (fissioned) into two nuclei, called fission products. Two or three neutrons are released with the fission products. If at least one of these neutrons produces fission in a neighboring uranium or plutonium nucleus, a self-sustaining fission chain reaction can be produced. This process is best achieved if the isotopes uranium-235 or plutonium-239 are used. These two isotopes are the key materials in any nuclear-weapon program. Each fission event produces energy. A fission chain reaction, involving a very large number of fission events, can therefore release a very large amount of energy. A significant nuclear explosion will only occur if there is a sufficient amount of uranium-235 or plutonium-239 present to support a self-sustaining fission chain reaction. The minimum amount of the material required for this purpose is called the critical mass.

A nuclear weapon produces a powerful explosion by releasing a very large amount of energy in a very short time

An amount somewhat larger than the critical mass, called a supercritical mass, is required to produce a fission chain reaction for a nuclear explosion. The larger the quantity of uranium-235 or plutonium-239 that is fissioned, the greater the explosive yield of the nuclear explosion. The nuclear-weapon designer's aim is to create a weapon that will not be blown apart until it has produced the size of explosion he requires. In other words, the aim is to

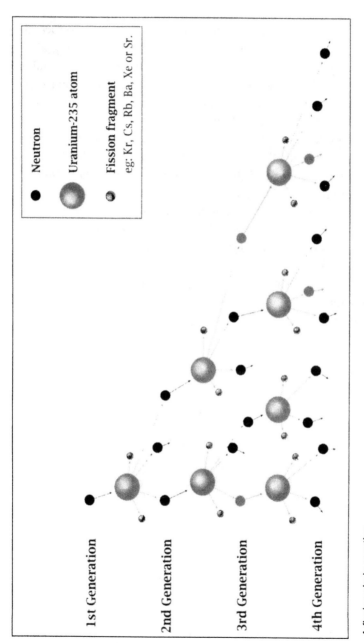

A fission chain reaction

keep the fission process going long enough to produce the required amount of energy. The most remarkable thing about nuclear weapons is the small amount of uranium-235 or plutonium -239 needed to produce a huge explosion: the critical mass of a sphere of plutonium-239 is about 11 kilograms; the radius of the sphere is only about 5 centimeters.

The plutonium sphere can be surrounded by a shell of a material like beryllium, the only function of which is to reflect back into the plutonium some of the neutrons that would otherwise have been lost to the fission chain reaction, increasing the number of fissions that take place. This trick reduces the critical mass considerably—typically, from 11 kilograms to about 4 kilograms, a sphere of a radius of approximately 3.6 centimeters, about the size of a small orange.

A fission nuclear weapon using just 4 kilograms of plutonium-239 would typically explode with a power of 20 kT, equivalent to that of the explosion of about 20,000 tons of TNT, the power of the

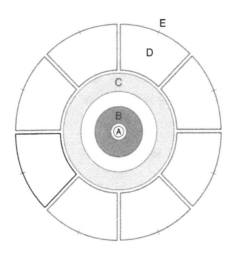

Ordinary nuclear fission weapon

Configuration of components of a fission bomb. A – initiator (neutron source or generator), B – fissile core (plutonium and U-235), C – tamper core reflector (uranium plus beryllium), D – high explosive lens (shaped plastic charge), E – detonator.

nuclear weapon that destroyed Nagasaki in August 1945. Such a nuclear weapon would therefore be about 5,000 times more effective, weight for weight, than a conventional bomb. The maximum explosive power of a militarily usable nuclear weapon using nuclear fission is about 50 kT because to obtain a bigger explosion a technique known as boosting is used. In a boosted weapon, some fusion material is injected into the centre of the plutonium mass as it is exploding with the result that the power of the explosion is boosted, typically tenfold.

Nuclear fusion occurs when nuclei of atoms of hydrogen isotopes fuse together to form nuclei of helium. Whereas fission involves the splitting of the nuclei of heavy isotopes like plutonium, fusion involves the joining together of light nuclei like hydrogen. Nuclear fusion takes place when the hydrogen nuclei are subject to very high temperatures and pressures, similar to those that occur in the sun, and exploding plutonium produces these conditions.

During fusion, neutrons are produced and energy is released. In a boosted weapon, these fusion neutrons are used to produce more fission in the plutonium-239. Boosted weapons are, therefore, sophisticated fission weapons. Tritium and deuterium, isotopes of hydrogen, are used as the fusion material in boosted weapons.

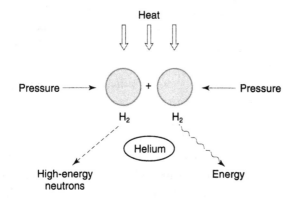

How nuclear fusion works

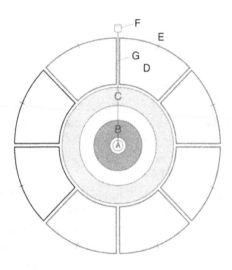

How a boosted weapon works

Configuration of components of a boosted fission bomb. A – initiator (neutron source or generator), B – fissile core (plutonium and U-235), C – tamper core reflector (uranium plus beryllium), D – high explosive lens (shaped plastic charge), E – detonator, F – tritium container, G – tritium feed into core of bomb.

Boosted weapons are very efficient, typically between five or ten times more than ordinary fission weapons. Much higher explosive powers can be obtained from a given amount of plutonium -239. Typically, explosive powers of up to 500 kT can be obtained from boosted fission weapons, ten times greater than the explosive powers that can be obtained from fission nuclear weapons that are not boosted. An explosion of this size would totally destroy a large city.

If even greater explosive powers than 500 kT are required, a large fraction of the energy must be obtained from nuclear fusion. Nuclear weapons that rely for their explosive power mainly on fusion are called thermonuclear weapons or H-bombs. In a thermonuclear weapon, a nuclear fission weapon acts as a trigger, providing the high temperature and pressure required for fusion. Typically, a cylinder of fusion material, in the form of lithium deuteride, is placed beneath the trigger. When the fission trigger explodes it generates fusion in the fusion stage.

There is no critical mass for the fusion process, and so in theory there is no limit to the explosive power that can be obtained from a thermonuclear weapon. In 1962, the former Soviet Union exploded a thermonuclear weapon at its Arctic test-site at Novaya Zemlya with an explosive yield equivalent to that of the explosion of nearly 60 million tons of TNT, or about 3,000 Nagasaki weapons. This is very much more explosive power than would be required to destroy totally the largest city on earth.

The nuclear powers

The Nuclear Non-Proliferation Treaty attempts to prevent the spread of nuclear weapons to countries other than China, France, Russia, the UK, and the USA (the five permanent members of the Security Council of the UN). The development, production, stockpiling, and use of both biological and chemical weapons are prohibited under international treaties. But these treaties have not prevented the proliferation of biological, chemical or nuclear weapons.

Nuclear weapons are the WMDs of choice of the five major powers mentioned above, and of three regional powers—India, Israel, and Pakistan. Iran and Iraq are strongly suspected of developing nuclear weapons and North Korea probably already has one or two.

There are about 30,000 nuclear weapons in today's world. Some are deployed in operational weapons; some are kept in reserve for possible future deployment; and some are waiting to be dismantled. The majority are American or Russian; the United States and Russia each deploy about 9,000 nuclear weapons. The other countries with nuclear weapons—China, France, the UK, India, Israel, and Pakistan—have a total of about 1,200 in their operational nuclear arsenals. China has deployed about 400 nuclear weapons, France about 350, and the UK about 200. Israel is estimated to have about 200, India about 60, and Pakistan about 35.

There are about 30,000 nuclear weapons in today's world

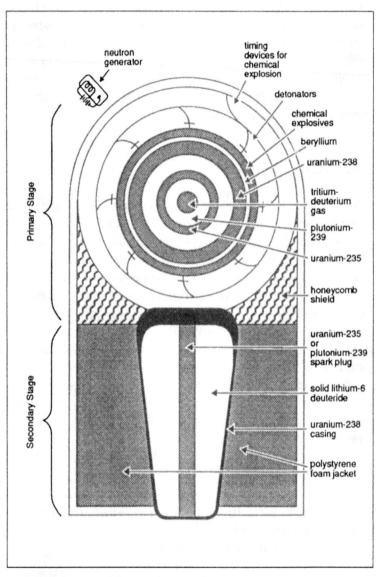

Schematic cross-section of a thermonuclear weapon

Nuclear weapons are far more destructive than conventional bombs. During the Second World War it took a number of raids, each involving 1,000 or more bombers, to destroy, for example, the German city of Dresden with high explosive and incendiary bombs, killing at least 50,000 people. Several times more people were killed in 1945 in each of the Japanese cities of Hiroshima and Nagasaki using a single nuclear weapon.

There are many types of nuclear weapons—aircraft bombs, artillery shells, depth charges, torpedoes, land mines, cruise missiles, and a variety of ballistic and other missiles. They may be deployed for tactical or strategic use. Tactical ones generally have a lower explosive yield and shorter range than strategic ones. The effect of the explosion of a nuclear weapon depends mainly on the explosive yield of the weapon and the altitude at which it explodes.

The explosive yields of the nuclear weapons currently deployed by the nuclear-weapon powers vary considerably. Some artillery shells, aircraft bombs, and land mines have low yields, some of less than 1 kT; strategic intercontinental and submarine-launched ballistic missiles have the highest yields. Some Chinese strategic nuclear weapons have yields as high as 5 megatons. All other deployed strategic nuclear weapons have yields of less than 750 kT.

What a nuclear explosion does

The author was present when British nuclear weapons were tested at Maralinga, in the South Australian desert, in 1953. Seeing a nuclear explosion is an awesome experience: the observer at first stands with his back to the explosion to avoid being blinded by the initial flash of light and ultraviolet radiation. After the flash, he can turn towards the nuclear explosion to watch the fireball grow.

The initial flash of light is followed by a weird, very short period of silence. Any exposed skin then feels a wave of heat. Just

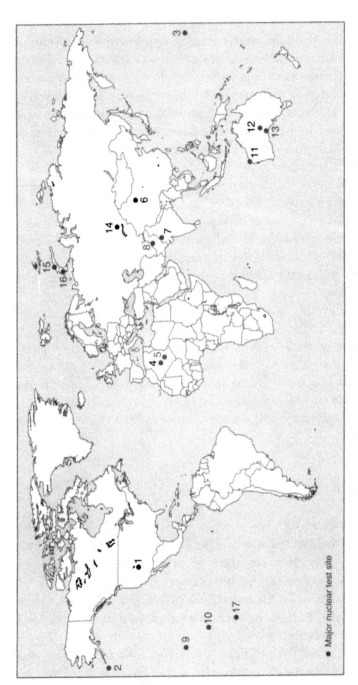

World map showing civil nuclear powers, date of first civil nuclear power reactor and major nuclear test sites worldwide

● Major nuclear test site

No.	Country	Site
1.	USA	Nevada test site (u and a)
2.	USA	Amchitca, Alaska (u)
3.	France	French Polynesia at Mororoa Atoll and Fangatuafa (u and a)
4.	France	Reggan, Sahara desert (a)
5.	France	Sahara desert (u)
6.	China	Lop Nor, Sinkiang Province (u and a)
7.	India	Thar desert, Jaisalmer district (u)
8.	Pakistan	Chagai Hills (u)
9.	UK	Pacific at Johnston Atoll (a)
10.	UK	Pacific at Christmas Island (a)
11.	UK	Monte Bello Islands (a)
12.	UK	Emu Field, South Australia (a)
13.	UK	Maralinga, South Australia (a)
14.	USSR	Novaya Zemlya, South Site (u and a)
15.	USSR	Semipalatinsk, Kazakhstan (u and a)
16.	USSR	Novaya Zemlya, North Site (u and a)
17.	USA	Pacific tests at Johnston Atoll, Enewetak, Bikini, Christmas Island (a)

u = nuclear tests performed underground
a = nuclear tests performed in the atmosphere

Year when countries commissioned their first nuclear-power reactors

Country	year of operation of first nuclear-power reactor	Country	year of operation of first nuclear-power reactor
Argentina	1974	Lithuania	1985
Armenia	1979	Mexico	1990
Belgium	1962	Netherlands	1969
Brazil	1985	Pakistan	1972
Bulgaria	1974	Romania	1996
Canada	1971	Russia	1954
China	1994	Slovakia	1973
Czeck Republic	1985	Slovenia	1983
Finland	1977	South Africa	1984
France	1964	Spain	1971
Germany	1966	Sweden	1972
Hungary	1983	Switzerland	1969
India	1969	Taiwan	1978
Italy	1964	UK	1956
Japan	1965	Ukraine	1978
Kazakhstan	1973	USA	1957
South Korea	1978		

Italy and Kazakhstan no longer have nuclear-power reactors.

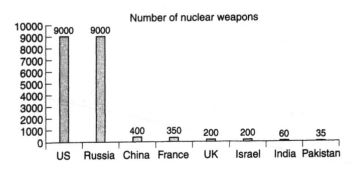

Nuclear arsenals by country

as one gets over the surprise of the heat wave, one is shaken by the blast wave, accompanied by a loud noise. The body is shaken again by a wind travelling away from the explosion, raising a cloud of dust. A short time later, one is shaken yet again by another wind blowing in the opposite direction.

Experiencing the heat, blast, noise, and the winds, seeing the brilliantly colored fireball growing to a tremendous size, and watching the mushroom cloud rise to a high altitude, combine to give a sense of the immense power of a single nuclear explosion. It is an experience that one does not forget. The most awesome thing is that this huge explosion, powerful enough to destroy a city, is produced by a piece of plutonium about the size of a tennis ball.

The initial flash of light is followed by a weird, very short period of silence

Nuclear weapons are quantitatively and qualitatively different from conventional weapons. Professor Sir Joseph Rotblat, in his book *Nuclear Radiation in Warfare*, explains:

> A single nuclear bomb can have an explosive yield greater than that of the total of all the explosives ever used in wars since gunpowder was invented. The qualitative difference which makes nuclear weapons unique is that, in addition to causing

Nuclear weapon exploding

loss of life by mechanical blast, or by burns from the heat of the fireball, nuclear weapons have a third killer—radiation. Moreover, and unlike the other two agents of death, the lethal action of radiation can stretch well beyond the war theatre and continue long after the war has ended, into future generations.

At the instant of the detonation of a typical nuclear weapon, the temperature shoots up to tens of millions of degrees and pressure to millions of atmospheres. As the fireball, a luminous mass of air, starts to expand, conditions are like those in the sun. The energy of the explosion is carried off by heat, blast and radiation. When a typical nuclear-fission weapon explodes, roughly half of the energy goes in blast, about a third in heat and the rest in radiation.

With a bomb of the size of the Hiroshima one, heat will kill people over a larger area than either blast or radiation. The lethal areas for blast and radiation are about the same; each is about half of the lethal area for heat. For weapons with much larger explosive yields, heat is by far the biggest killer, several times more lethal than either blast or radiation.

Heat, blast and radiation

In the first few thousandths of a second after the explosion begins there is a burst of ultraviolet radiation from the fireball as it rises in the atmosphere. This is followed by a second burst of radiation—thermal radiation—lasting for a few seconds. After a minute or so, the temperature of the fireball has fallen sufficiently for it to stop emitting visible light. The second burst of thermal radiation is responsible for the heat effects of the weapon. Exposed people will be killed or severely burned and fires will be started over a large area. The area affected will depend on the explosive yield of the weapon and the weather. If the weather is fine, the heat wave can kill and injure people at much greater distances than can blast and radiation.

About half of the people caught by the heat wave at a distance of 2 kilometers from the explosion of a nuclear weapon with a yield of 12 kT (similar to the atomic bomb that destroyed Hiroshima) at low altitude in fine weather will suffer third-degree burns. For a nuclear explosion of 300 kT, the distance would be about 7.5 kilometers. People will also be killed and injured by fires set alight by the thermal radiation.

If a nuclear weapon is exploded over a town or city, most immediate death and injury will be caused by blast. Blast will also be the main cause of damage to buildings. In fact, most blast deaths occur from indirect effects—falling buildings and debris, being hurled into objects by the blast wave, and so on. For a 12 kT bomb exploded at a height of 300 meters, the lethal blast area is about 5 square kilometers.

During the first minute following a nuclear explosion, ionizing radiation, called initial radiation, is given off. Ionizing radiation emitted after a minute is called residual radiation, most of which comes from the fallout of radioactive fission products. Much of the radioactivity in the fallout will be in the mushroom cloud produced by the explosion.

As the cloud is blown downwind, radioactive particles will fall to the ground. People in the contaminated area may then be exposed to the radiation given off by the radioactive particles.

They can be irradiated by radiation given off by radioactive fallout; they can also inhale or swallow radioactivity from fallout and then be irradiated by the radioactivity in their bodies.

Radiation causes atoms to become electrically charged, a process called ionization. Cells in the body consist of atoms, and if one of the atoms in a cell is ionized it can be dangerous. When a person is exposed to low levels of radiation, cells will be damaged but the body can repair this damage, but when the body is exposed to higher doses of radiation so many cells are damaged that the body's repair mechanisms cannot cope.

when the body is exposed to higher doses of radiation so many cells are damaged that the body's repair mechanisms cannot cope

Some cells are more easily damaged by radiation than others. The most sensitive cells are those that line the intestines, white

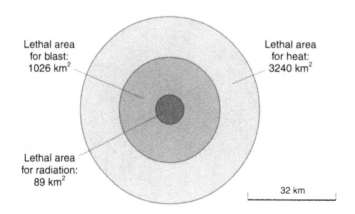

Lethal area of a 60mT bomb

("Lethal area" is defined as the area in which the number of survivors equals the number of fatalities outside the area)

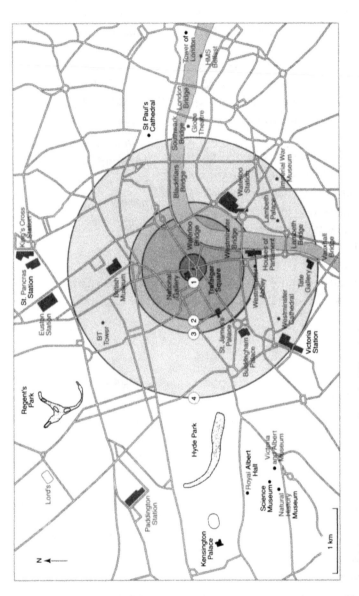

Map of central London showing effect of 1 kT explosion

Within Circle 1 (200 meters), almost 100 percent fatality in those directly exposed to thermal radiation; within Circle 2 (800 metres), almost 100 percent fatality in those directly exposed to blast; within Circle 3 (one kilometer), almost 100 percent fatality in those directly exposed to prompt nuclear radiation; within Circle 4 (two kilometers), almost all directly exposed suffer immediate injuries from burns and blast. (Map by Antony Smith with additional graphics by Richard Prime.)

blood cells that combat infection, and cells that produce red and white blood cells. The effects of radiation on these types of cells lead to the first symptoms of radiation sickness, including nausea, diarrhea, vomiting and fatigue. These symptoms may be followed by, among others, headache, hair loss, dehydration, breathlessness, hemorhage, anemia, permanent darkening of the skin, loss of weight, fever, fatigue and sweating. All of these symptoms occur only at high doses of radiation; with lower doses only some of them may occur.

Very high doses of ionizing radiation can produce symptoms within minutes. Death may occur from short-term (acute) effects within about two months. Death from long-term effects, particularly leukemia, may occur several years later and other cancers may occur after very long times, of thirty or more years.

The reality of nuclear attack—eyewitness accounts

8:15 a.m.—atomic bomb released—43 seconds later, a flash— shock wave, craft careens—huge atomic cloud

9:00 a.m.—cloud in sight—altitude more than 12,000 meters

Part of the flight log of the *Enola Gay*, the American B-29 bomber that atom-bombed Hiroshima, August 6, 1945.

The pilot's story

The *Enola Gay* dropped the bomb that destroyed Hiroshima from an altitude of about 7,900 meters; the bomb exploded at an altitude of 570 meters. Paul Tibbets, the pilot of the *Enola Gay*, explained that he told his air crew that he would say, as the *Enola Gay* approached Hiroshima,

"One minute out," "Thirty seconds out," "Twenty seconds," and "Ten" and then I'd count, "Nine, eight, seven, six, five, four seconds," which would give them time to drop their cargo (the atomic bomb). They knew what was going on because they knew where we were. And that's exactly the way it worked, it was absolutely perfect. We get to that point where I say "one second" and by the time I'd got that second out of my mouth the airplane had lurched, because 10,000 l.b.s. (the weight of the bomb) had come out of the front. I'm in this turn now, tight as I can get it, that helps me hold my altitude and helps me hold my airspeed and everything else all the way round. When I level out, the nose is a little bit high and as I look up there the whole sky is lit up in the prettiest blues and pinks I've ever seen in my life. It was just great.

I tell people I tasted it. "Well," they say, "what do you mean?" When I was a child, if you had a cavity in your tooth the dentist put some mixture of some cotton or whatever it was and lead into your teeth and pounded them in with a hammer. I learned that if I had a spoon of ice-

the whole sky is lit up in the prettiest blues and pinks I've ever seen in my life

cream and touched one of those teeth I got this electrolysis and I got the taste of lead out of it. And I knew right away what it was. OK, we're all going. We had been briefed to stay off the radios: "Don't say a damn word, what we do is we make this turn, we're going to get out of here as fast as we can." I want to get out over the sea of Japan because I know they can't find me over there. With that done we're home free.

The shockwave was coming up at us after we turned. And the tailgunner said, "Here it comes." About the time he said that, we got this kick in the ass. I had accelerometers installed in all airplanes to record the magnitude of the bomb. Next day, when we got figures from the scientists on what they had learned from all the things, they said, "When that bomb exploded, your airplane was ten and a half miles away from it."

You see all kinds of mushroom clouds, but they were made with different types of bombs. The Hiroshima bomb did not make a mushroom. It was what I call a stringer. It just came up. It was black as hell, and it had light and colors and white in it and grey color in it and the top was like a folded-up Christmas tree.

A survivor's story

Scientific descriptions of the effects of the explosion of a nuclear weapon over a city cannot convey the awesome power of a nuclear explosion nearly as well as eyewitness accounts. The difference is dramatically brought home by the eloquence of the accounts of survivors of the nuclear destruction of Hiroshima and Nagasaki.

Kataoka Osamu was a teenage schoolboy in Hiroshima when the bomb exploded. His moving account of his experience is in stark contrast to the detached, matter-of-fact account of the pilot.

"I looked out of the window at the branch of a willow tree," he remembers.

Just at the moment I turned my eyes back into the old and dark classroom, there was a flash. It was as if a monstrous piece of celluloid had flared up all at once. Even as my eyes were being pierced by the sharp vermilion flash, the school building was already crumbling. I felt plaster and roof tiles and lumber come crashing down on my head, shoulders and back. The dusty smell of the plaster and other strange smells mixed up with it penetrated my nostrils.

I wondered how much time had passed. It had gradually become harder and harder for me to breathe. The smell had become intense. It was the smell that made it so hard to breathe.

I was trapped under the wreckage of the school building . . . I finally managed to get out from under the wreckage and stepped out into the schoolyard. It was just as dark outside as it had been under the wreckage and the sharp odor was everywhere. I took my handkerchief, wet it, and covered my mouth with it.

Four of my classmates came crawling out from beneath the wreckage just as I had done. In a daze we gathered around the willow tree, which was now leaning over. Then we began singing the school song. Our voices were low and rasping, with a tone of deep sadness. But our singing was drowned out by the roar of the swirling smoke and dust and the sound of the crumbling buildings.

We went to the swimming pool, helping a classmate whose leg had been injured and who had lost his eyesight. You cannot imagine what I saw there. One of our classmates had fallen into the pool; he was already dead, his entire body burned and

Hiroshima bomb damage

tattered. Another was trying to extinguish the flames rising from his friend's clothes with the blood which spurted out of his own wounds. Some jumped into the swimming pool to extinguish their burning clothes, only to drown because their terribly burned limbs were useless. There were others with burns all over their bodies whose faces were swollen to two or three times their normal size so they were no longer recognizable. I cannot forget the sight of those who could not move at all, who simply looked up at the sky, saying over and over, "Damn you! Damn you!"

> *I cannot forget the sight of those who could not move at all, who simply looked up at the sky, saying over and over, "Damn you! Damn you!"*

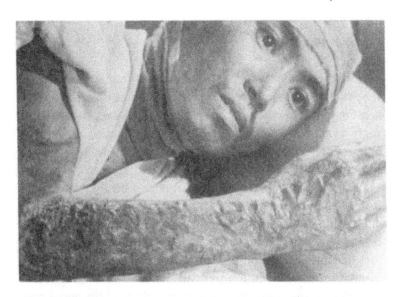

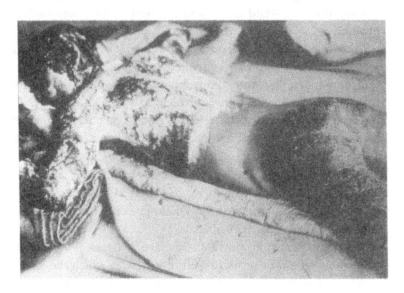

Hiroshima victims

Nuclear terrorism

Terrorists would have to obtain suitable uranium or plutonium to fabricate a crude nuclear explosive. They are more likely to acquire plutonium than uranium because it is becoming increasingly available (see Chapter 7). Civil plutonium is separated from spent civil nuclear-power reactor fuel in reprocessing plants, such as those operated at Sellafield, England; La Hague, France; and Chelyabinsk, Russia. Another is being constructed at Rokkashomura, Japan.

A group of two or three people with appropriate skills could design and fabricate a crude nuclear explosive. It is a sobering fact that the fabrication of a primitive nuclear explosive using plutonium or suitable uranium would require no greater skill than that required for the production and use of the nerve agent produced by the AUM group and released in the Tokyo underground.

A crude nuclear explosive designed and built by terrorists could well explode with a power equivalent to that of 100 tons of TNT. For comparison, the largest conventional bombs used in warfare so far had explosive powers equivalent to about 10 tons of TNT. The terrorist bomb set off at the World Trade Center in 1993 had an explosive power equivalent to that of about a ton of TNT, the one that destroyed the Murrah building in Oklahoma in 1995 that of about 2 tons of TNT, and the one that destroyed the Al Khobar Towers building near Dhahran, Saudi Arabia, in 1996 that of about 4 tons of TNT. The size of the Dhahran bomb surprised and shocked American security officials.

A nuclear explosion equivalent to that of 100 tons of TNT in an urban area would be a catastrophic event, with which the emergency services would be unable to cope effectively. Exploded on or near the ground, it would produce a crater, in dry soil or dry soft rock, about 30 meters across. The area of lethal damage from the blast would be roughly 0.4 square kilometers; the lethal area for heat would be about 0.1 square kilometers.

The direct effects of radiation, blast or heat would very probably kill people in the open within 600 meters of the explosion. Many other deaths would occur, particularly from indirect blast effects such as the collapse of buildings.

Heat and blast will cause fires, from broken gas pipes, gasoline in cars, and so on. The area and extent of damage from fires may well exceed those from the direct effects of heat.

The area significantly contaminated with radioactive fallout will be uninhabitable until decontaminated. It may be many square kilometers and it is likely to take a long time to decontaminate it to a level sufficiently free of radioactivity to be acceptable to the public.

An explosion of this size, involving many hundreds of deaths and injuries, would paralyze the emergency services. They would find it difficult even to deal effectively with the dead. Many, if not most, of the seriously injured would die from lack of medical care. In the UK, for example, there are only a few hundred burn beds in the whole country.

There would be considerable delays in releasing injured people trapped in buildings. And, even for those not trapped, it would take a significant time to get ambulances through to them and then to transport them to the hospital. A high proportion of the seriously injured would not get medical attention in time to save them. This scenario of a nuclear terrorist attack would put a far greater strain on the emergency services than did the attack on New York on September 11, 2001.

The simplest and most primitive terrorist nuclear device would be a radiological weapon or radiological dispersal device, commonly called a "dirty bomb". It is not strictly speaking a nuclear weapon, as it does not involve a nuclear explosion. A dirty bomb would

There are literally millions of radioactive sources used worldwide in medicine, industry and agriculture; many of them could be used to fabricate a dirty bomb

consist of a conventional high explosive—for example, Semtex, dynamite or TNT—and a quantity of a radioactive material. Incendiary material, such as thermite, is likely to be put into a dirty bomb to produce a fierce fire when the bomb is set off. The radioactivity would be taken up into the atmosphere by the fireball and would then be blown downwind.

There are literally millions of radioactive sources used worldwide in medicine, industry, and agriculture; few of them are kept securely and many of them could be used to fabricate a dirty bomb. The most likely to be used are those that are relatively easily available, have a relatively long half-life, of several months or years, and emit energetic gamma radiation; suitable candidates include caesium-137, cobalt-60, and strontium-90.

Clearly, the lack of security on radioactive materials around the world is a major cause for concern; even in the United States and Europe, where security is comparatively strong, there are thousands of instances of radioactive sources that have been lost or stolen over the years. Their present whereabouts are unknown.

Effects of a radiological weapon

The detonation of a dirty bomb is unlikely to cause a large number of casualties. Generally, any immediate deaths or serious injuries would most likely be caused by the detonation of the conventional explosive. The radioactive material in the bomb would be dispersed into the air but would soon be diluted to relatively low concentrations.

If the bomb were exploded in a city, as it almost certainly would be, some people would probably be exposed to a dose of radiation. But in most cases the dose would probably be relatively small. A low-level exposure to radiation would slightly increase the long-term risk of cancer.

The main potential impact of a dirty bomb is psychological—it would cause considerable fear, panic, and social disruption, exactly the effects terrorists wish to achieve. The public fear of radiation is very great indeed, some say irrationally or disproportionately so.

The radioactive area would have to be evacuated as quickly as possible, to prevent people becoming contaminated, and would then have to be decontaminated. The degree of contamination would depend on the amounts of high explosive and incendiary material used, the amount and type of radioisotope in the bomb, whether it was exploded inside a building or outside, and the weather conditions. Decontamination is likely to be very costly (costing millions of dollars) and take weeks or, most likely, many months to complete. Radioactive contamination is the most threatening aspect of a dirty bomb.

2 Biological weapons

What is a biological weapon?

Biological weapons spread disease deliberately in human populations, when people are exposed to infectious microorganisms or to the toxins they produce. They may also affect animals and plants. Biological weapons can be much more lethal than chemical weapons but are less so than the most powerful nuclear weapons.

According to an official American study, about 30 kilograms of anthrax spores could kill more people than the nuclear weapon that destroyed Hiroshima (equivalent to 12,500 tons of TNT). An estimate of the number of fatalities from the nuclear weapon would be between 23,000 and 80,000 people, whereas the anthrax could kill between 30,000 and 100,000.

Biological weapons are by no means new. The use of human or animal corpses to befoul wells is the most ancient use of biological warfare, instances of which are recorded in early Persian, Greek, and Roman literature. Examples of the use of corpses to contaminate drinking water occur up to the twentieth century in European wars, the American Civil War, and the South African Boer War.

In 1346, the town of Feodosia, in the Crimea, a Genoese trading outpost, was withstanding a siege by a Tartar army. When the besieging Tartars were struck by a plague epidemic they catapulted their plague victims into the town where the disease rapidly spread. The survivors were forced to flee back to Italy by sea, taking the plague with them. In 1710, in the battle against Swedish troops in Reval, the Russian besiegers threw bodies of plague victims over the city walls, causing an epidemic, and in 1763, the British killed North American Indians by making them presents of hospital blankets taken from smallpox patients. Two hostile Indian tribes were given two blankets and a handkerchief taken from the smallpox hospital. The stratagem worked only too well. Within a few months, smallpox was prevalent among the various Indian tribes in the Ohio region and these peoples were decimated. But the ruse backfired: the Americans also used smallpox against the British during the Revolutionary War.

Despite these historical precedents, biological weapons have not been used to any significant extent in modern times. The British actively considered using anthrax against Germany in the Second World War, but both then and since, fear of retaliation has prevented countries from engaging in biological warfare.

What are biological-warfare agents?

There are four types of biological-warfare agents, all disease-carrying substances (toxins) or else microorganisms—bacteria, viruses, rickettsiae, and fungi. Bacteria, single-cell microorganisms, cause such diseases as anthrax, cholera, pneumonic plague, and typhoid. They produce illnesses by invading tissues and/or by producing poisonous toxins including botulinum, ricin, tetanus, and diphtheria.

Viruses, the simplest form of organisms, cause diseases such as Ebola, AIDS, flu, polio, and smallpox. They cannot live independently and must, therefore, invade living cells to reproduce and grow.

Rickettsiae are microorganisms, intermediate between bacteria and viruses, found in the tissues of lice, ticks, and fleas, and can cause diseases such as typhus, Q-fever, and Rocky Mountain spotted fever when transmitted to humans. Fungi, more complex organisms than bacteria, reproduce by forming spores and cause diseases like coccidiomycosis.

Microorganisms such as those produced by *Clostridium botulinum*, which causes botulism, produce toxins. Whereas bacteria and viruses can spread disease through a population by contact between humans, toxins cannot.

There are a bewildering number of possible biological-warfare agents. The ones most likely to be deployed are: anthrax, outline toxin, smallpox, and ricin. Iraq, for example, had produced by 1991 significant quantities of anthrax and of botulinum toxin. Any Iraqi biological weapons found by UNSCOM, the United Nations teams of inspectors, were destroyed. The inspectors left Iraq in December 1998 and returned in 2002.

Anthrax

Inhalation anthrax, the type of the disease most likely to be used in biological warfare, occurs when bacteria *Bacillus anthracis* are breathed into the lungs. The disease is usually not diagnosed in time for treatment and the mortality rate is typically about 95 percent. The first symptoms of inhalation anthrax are nonspecific, including fever, malaise, and fatigue, sometimes with a dry hacking cough. After these symptoms occur, treatment cannot help. After about three days, severe respiratory distress occurs and death usually follows within thirty-six hours.

Anthrax is a preferred biological-warfare agent because it is highly lethal, it is very stable and it is relatively easy to disperse

Anthrax is a preferred biological-warfare agent of states such as Iraq because it is highly lethal, it is easily produced in large quantities at low cost, it is very stable and can be stored for a very long period as a dry powder, and it is relatively easy to disperse as an aerosol with crude sprayers. For the same reasons, anthrax is also likely to be the preferred biological agent for terrorists.

Botulinum toxin

Botulinum toxin is extremely toxic. Symptoms of inhalation botulism may begin within thirty-six hours after exposure or they may be delayed for several days. The first symptoms include general weakness, dizziness, extreme dryness of the mouth and throat, the retention of urine, blurred vision and sensitivity to light. Respiratory failure caused by paralysis of respiratory muscles is generally the cause of death. Fatalities may be limited to no more than 5 percent with effective treatment but intensive and prolonged nursing care may be required for recovery. When ingested, botulinum toxin can cause food poisoning.

Smallpox

Smallpox is a highly contagious disease caused by the virus variola. One strain has a high mortality of about 30 percent. A global campaign, begun in 1967 and administered by the World Health Organization, eradicated smallpox by 1977. The WHO recommended that all stocks of smallpox virus should be destroyed or transferred to one of two laboratories—at the Institute of Virus Preparations in Moscow, or at the Centers for Disease Control and Prevention (CDC) in Atlanta, Georgia. All countries claimed to have complied.

However, there are fears that other countries may since have illegally acquired some smallpox viruses from sources in countries that did not send all their stocks of virus to the designated laboratories or from people with access to the stocks held in the designated laboratories. The most likely source is the Moscow laboratory. Russian scientists and technicians are so badly paid that the temptation to steal and sell the smallpox virus must be considerable. If any country has acquired viruses, smallpox may be used in the future as a biological weapon.

The smallpox virus is very stable in aerosol form and its release could infect a large number of people, spreading from person to person by droplets or direct contact. After an incubation period of seven to seventeen days, the patient experiences high fever, headache, backache, vomiting, prostration, and possibly delirium. After another two or three days the smallpox rash appears, turning into turbid blisters after about five days. Death, which typically occurs during the second week of the illness, is usually caused by toxemia (blood poisoning).

Ebola hemorrhagic fever

The biological agents described above are bad enough, but a frightening possibility is that the virus that causes Ebola hemorrhagic fever will be used as a biological-warfare agent. Ebola is a truly horrible disease, one of the most lethal, fatal in little more than a week in up to 90 percent of those infected. Such lethality makes Ebola virus a potentially attractive biological-warfare agent.

The Ebola virus, found in the rainforests of Africa and Asia, is transmitted by direct contact with the blood, secretions, organs, or semen of infected people. After an incubation period of between two and twenty-one days, the symptoms of Ebola are the sudden onset of fever, weakness, muscle pain, headache, and sore throat. A rash, vomiting, diarrhea, reduced kidney and liver function, and convulsions follow. Both internal and external bleeding occurs—every orifice bleeds. Convulsive victims splash infected blood around them, shaking and thrashing as they die. No cure exists.

> *Both internal and external bleeding occurs—every orifice bleeds*

Ricin

Ricin is a toxin obtained from the seeds of the castor oil plant *Ricinus communis*. Ricin blocks the synthesis of proteins in the cells of the body, killing the cell. Ricin is a likely biological-warfare agent because it is easy to produce, and has a very high inhalation toxicity. Ricin, reportedly obtained from the KGB, was used to assassinate the exiled Bulgarian broadcaster Georgi Markov while he was waiting for a bus in London. The agent was injected into his lower leg using the sharpened end of an umbrella. He developed severe gastroenteritis and died three days later.

If ingested, ricin causes a rapid onset of nausea, vomiting, abdominal cramps and severe diarrhea with vascular collapse. Death may occur from the third day. If inhaled, ricin may cause weakness, fever, coughing, and hypothermia followed by hypotension and cardiovascular collapse. High doses by inhalation may produce sufficient damage to the lungs to cause death. Currently, no antitoxin or prophylactic treatment for ricin poisoning exists.

Foot-and-mouth disease virus

A terrorist attack with the virus that causes foot-and-mouth disease in animals, the most infectious virus known, is particularly feared. The outbreak of this gruesome disease in the United Kingdom in 2001, with nightly television pictures of enormous pyres of burning animal carcasses, dramatically demonstrated the terrible consequences of the disease.

Foot-and-mouth disease is an acute, highly contagious viral infection of cloven-hoofed animals. There are seven main types, all with similar symptoms. Even when animals recover from infection by one type of virus they have little or no protection against attacks by any one of the others.

The disease is present in many countries of the world, and is endemic in parts of Asia, Africa, the Middle East, and South America, with sporadic outbreaks in disease-free areas. North and Central America, Australia, New Zealand, the United Kingdom, and Scandinavia do not have the disease. In, for example, 2001 there were outbreaks in Butan, Brazil, Columbia, Egypt, Georgia, Japan, Kazakhstan, Korea, Kuwait, Malawi, Malaysia, Mongolia, Namibia, Russia, South Africa, Taipei, Tajikstan, Uruguay, and

Zambia. Terrorists would, therefore, have no difficulty in acquiring supplies of the virus.

The disease is highly contagious and may spread over great distances with movement of infected or contaminated animals, products, objects, and people; airborne spread of the disease can also take place. Floyd P. Horn and Roger G. Breeze explain: "It can spread over 170 miles as an aerosol on the wind from an infected farm. One infected pig releases enough virus each day to infect, theoretically, 100 million cattle."

Cattle, sheep, pigs, and goats are susceptible to foot-and-mouth disease and so are some wild animals such as hedgehogs, coypu, rats, dee,r and zoo animals including elephants. Pigs are mainly infected by ingesting infected food. Cattle are mainly infected by inhalation, often from pigs, which excrete large amounts of virus by respiratory aerosols and are very important in spreading the disease. Large amounts of virus are excreted by infected animals before clinical signs are evident. People can be infected through skin wounds, by inhalation while handling diseased animals, by drinking infected milk, but not by eating meat from infected animals. The human infection is temporary and mild. The incubation period is between two and twenty-one days although the virus can be spread before clinical signs develop.

The rate of infection in animals can reach 100 percent, and mortality can range from 5 percent in adult animals to 75 percent in young pigs and sheep. Recovered cattle may be carriers for eighteen to twenty-four months; sheep for one to two months. Pigs are not carriers. Clinical signs in cattle are salivation; depression; anorexia; lameness caused by the presence of painful blisters, chiefly in the mouth or on the feet, but also in the skin of the lips, tongue, gums, nostrils, and teats; fever; decreased milk production; abortion and sterility. Lameness is the predominant symptom.

Because of the range of species affected, the high rate of infectivity, and the fact that virus is shed before clinical signs occur, foot, and mouth disease is one of the most feared animal diseases, capable of costing billions of pounds in lost production, loss of export markets, and loss of animals during eradication of the disease. There is no cure for foot-and mouth-disease. It usually runs its course in two or three weeks after which the great majority of animals recover naturally. Animals are slaughtered because widespread disease throughout the country would be economically disastrous. The economic and social consequences of foot-and-mouth disease and the terrible effects on the morale of the population make the disease an attractive terrorist weapon.

The economic and social consequences of foot-and-mouth disease make the disease an attractive terrorist weapon

The making of biological weapons

The facilities and resources needed to produce and deploy biological weapons are much less than those required for a nuclear-weapon program. The main reason why biological weapons are cheaper to produce than nuclear and chemical weapons is that the organisms used in the weapons reproduce themselves so that only a small quantity of them are needed to start the production process. The ability to reproduce is a unique characteristic of biological-warfare agents unlike chemical or nuclear weaponry.

The first step in the production of biological weapons is the selection of the organism or organisms to be used in the weapons. A small culture of the chosen organism is then acquired and used for the large-scale production of the organisms in a suitable facility. The organisms will be treated to prevent them degrading and then stored until loaded into a biological munition. Stabilization is also required to minimize degradation when the organisms are dispersed by the munition. The final step in a biological-warfare program is the choice of munitions and their production.

A military facility for the production of biological-warfare agents is very similar in design to a civilian facility to produce antibiotics or vaccines of a pharmaceutical grade, and it contains the same sort of equipment. Both facilities will employ microbiologists, physical chemists and biophysicists to advise on methods of freeze-drying and so on, and trained technicians.

Often the same toxic agent is required for both biological and toxin warfare and vaccine production. "Initial processing of agents and processing of their associated vaccines may differ only by a few steps, and then often only in the degree of care taken and subtle differences in method (such as the degree of purification and the type of containment used)," David Isenberg, analyst of biological technology equipment, explains. "The facilities required for the production of biological and toxin warfare agents are the same as those used in legitimate vaccine or pharmaceutical plants. Both include equipment and materials for microbial fermentation, cell culture and egg incubation, followed by harvesting and purification."

The facility will include measures to contain the biological agent to prevent the workers becoming exposed to it and infected, including the use of barriers to separate workers from the agent and highly efficient methods of ventilation. Strict containment is not necessary in, for example, a civilian antibiotic factory; on the contrary, in the civilian facility the product must be protected against contamination by materials in the environment.

Another difference is the degree of purification necessary. In a civilian biopharmaceutical establishment the products must be prepared to very high standards of purity to ensure that materials in the environment do not contaminate them. The operations are, therefore, carried out in sterile conditions. A military facility, however, does not have to achieve high levels of purity. On the contrary, a very pure biological-warfare agent is generally less stable than an unpurified one.

Sterilization is an important activity in a biological production facility. Equipment is sterilized, normally using saturated steam under pressure, before the product is processed. David Isenberg adds:

Equally important is the removal of condensate formed on the equipment. This prevents the formation of pockets of standing water, which promote bacterial growth, and maintains the high temperature necessary for sterilization. Supplying sterile, inert gases to processing equipment is a method of containment. This can protect oxygen-sensitive biomaterials and prevent aerosol generation of toxic products. Inert gases, such as nitrogen, helium, and argon, are usually supplied directly to processing equipment through sterile, in-line filters, maintaining a pressurized system or providing an inert blanket over the product in processing vessels.

Biological-warfare agents created by fermentation, or in tissue cultures or chick embryos, which include bacteria, toxins produced by bacteria, viruses, and rickettsiae, are normally colloidal liquids. (In a colloid, like milk or paint, the solid material is uniformly suspended in liquid.) The best way of storing cultures of bacteria is to freeze-dry the liquid. Freeze-drying is a technique for freezing cultures sufficiently rapidly to prevent the formation of ice crystals, followed by dehydration in a vacuum. The dried biological-warfare agents look like talcum powder.

When required for use, freeze-dried cultures can be simply rehydrated and cultured as usual. Organisms that cannot be effectively freeze-dried can be prepared for storage by ultra-freezing: storage of the contained material in, for example, liquid nitrogen refrigerators or very low temperature mechanical refrigerators. A toxin agent is usually prepared as a freeze-dried powder and then stored before loading into the munition.

Bacteria and bacterial toxins are grown in fermenters. Viruses, however, will not reproduce outside a living cell and must, therefore, be replicated in chick embryos (that is, in eggs), or in tissue cultures. Rickettsiae are replicated in the same way.

Fermenters are the most important pieces of equipment in a plant producing biological-warfare agents. Also crucial is equipment for freeze-drying.

Normally made of stainless steel, but sometimes of glass or plastic, fermenters are used to culture cells under carefully controlled conditions—such as temperature, acidity, and supply of nutrients—specifically chosen to suit the type of organism being cultured. A typical fermenter is cylindrical in shape, about 1 to 1.5 meters in diameter and 1 to 2 meters in height, fitted with ports through which nutrients can be introduced and samples of contents removed.

The plant will also contain high-speed centrifuges to separate the components of organisms, and autoclaves to sterilize, using steam at high pressure, the growth media and decontaminate equipment after use.

Much of the equipment used in a military plant making biological-warfare agents is dual purpose, very similar to that used in, for example, a brewery for making beer. It can therefore be bought from commercial suppliers without raising suspicions.

How biological-warfare agents are spread

If a biological weapon is to be effective the agent must be effectively dispersed over the target, generally as an aerosol, a cloud of very small droplets. An aerosol will remain airborne for some time; thus, as the cloud is blown along by the wind, the agent will fall to the ground slowly but steadily, contaminating the area on which it falls.

Aerosol technologies have been extensively developed for a number of civilian applications, among them the agricultural dispersal of pesticides and for spraying paint. These can be readily modified for military or terrorist use. The liquid agent is either forced under pressure through a fine nozzle, a technique called hydraulic atomization, or allowed to flow in a fine stream into a current of air, so-called air-blast atomization.

Biological-warfare munitions include free-fall aircraft bombs, artillery shells and rockets, sprayers carried in helicopters and aircraft, unmanned aircraft (remotely piloted aircraft), bomblets and cluster weapons, and ballistic missiles. Ballistic missiles are the preferred munitions for the strategic use of biological weapons.

A simple but effective way of dispersing a liquid biological-warfare agent is an aircraft spray tank. The agent is allowed to flow into, or just below, the slipstream of the aircraft where it is converted into small drops of a suitable size. Terrorists could acquire an aircraft with a spray tank, normally used for agricultural purposes, to disperse a biological agent.

Powdered biological agents can be dispersed using a small cylinder of compressed air, arranged to direct a stream of air over the surface of the powder, blowing it out uniformly through an exit slot.

What biological-warfare agents do

Graham S. Pearson, former Director General of the British Chemical and Biological Defence Establishment at Porton Down, Wiltshire, argues that, in some circumstances, the effects of the use of biological agents can be comparable to those of the use of a nuclear weapon. In general, however, nuclear weapons are far more lethal than biological weapons and, unlike nuclear weapons, biological weapons do not destroy or damage structures. Moreover, biological weapons kill and injure people over smaller areas than nuclear weapons.

Ballistic missiles are the preferred munitions for the strategic use of biological weapons

And those who survive a nuclear explosion are more likely to suffer significant psychological consequences including fear of leukemia and of cancer, mental disorder and possibly suicide.

Biological terrorism

The anthrax attacks in the United States following the September 11 terrorist attacks on New York and Washington, which killed five people, nonfatally infected seventeen others and caused considerable economic damage, greatly increased the fear of biological terrorism.

Terrorists could acquire biological agents from, for example, medical research laboratories by theft or from a sympathizer working there. Alternatively, biological agents could be obtained from natural sources. For example, the bacterium *Clostridium botulinum* is present in soil. A small sample could be cultured to provide large amounts for the large-scale contamination of food, possibly in food-processing factories. Infected individuals would become very ill; some would die. Anthrax, brucellosis and plague could also be acquired from natural sources. The spores of anthrax, which survive for decades, could be collected from soil in areas where anthrax is endemic in cattle, as could brucellosis bacteria.

Ease of acquisition is one reason why terrorists are likely to find biological agents attractive. They are also cheap. Biological agents are relatively easily dispersed: a slurry of anthrax spores could, for example, be prepared and deposited in underground train tunnels. When dry, the spores would be swept through the tunnel system by passing trains. Large numbers of people would be killed. Biological agents could also be freeze-dried and later rehydrated and dispersed as an aerosol.

3   Chemical weapons

What is a chemical weapon?

Chemical weapons are designed for the effective dispersal of a chemical-warfare agent. They were first used on a significant scale by both sides in the First World War, causing about 1,300,000 casualties, including about 90,000 deaths. Chemical weapons were used by Italy during its invasion of Ethiopia, 1935–6, Japan during its war against China, 1937–43, by Egypt against the Yemen, 1963–8, and by the United States in Vietnam, 1965–75. They were used by both sides in the Iran–Iraq war, 1980–88 and by Iraq against the Kurds at Halabja in 1988.

There are four main categories of chemical-warfare agents—choking, blister, blood and nerve agents.

Choking agents

Choking agents, such as carbonyl chloride or phosgene, attack the respiratory tract making the membranes swell and the lungs fill with fluid so that the victim drowns. Survivors normally suffer from chronic breathing problems.

Choking agents attack the respiratory tract making the membranes swell and the lungs fill with fluid so that the victim drowns

Blister agents

The best-known blister agent is mustard gas—bis(2-chloroethyl) sulphide, also called Yperite, a persistent agent, which remains toxic for a long period and can be lethal.

Mustard gas was extensively used by both sides in the First World War; the Germans, for example, fired more than a million shells, containing 2,500 tons of mustard gas, in the ten days following its introduction in 1917. The use of mustard gas during that conflict, and pictures of large numbers of blind soldiers, made it notorious. The public found the effects of chemical weapons so odious that political leaders in Europe and America were persuaded to negotiate the Geneva Protocol, signed on June 17, 1925, banning the use in war of "Asphyxiating, Poisonous or Other Gases, and of Bacteriological Methods of Warfare." More recently, both sides used mustard gas during the 1980–88 Iran–Iraq war, though both were parties to the Geneva Protocol.

Blood agents

Blood agents such as hydrogen cyanide and cyanogen chloride are absorbed into the body by breathing and kill by entering the bloodstream.

Nerve agents

There are two main groups of nerve agents—the G-agents, typically volatile liquids, that break down relatively quickly in the environment and cause death mainly when inhaled, and the V-agents, which are much more persistent and can be absorbed through the skin.

The most lethal nerve agents are three G-agents—tabun, soman, and sarin—and a V-agent (VX). VX, which typifies V-agents, is more persistent and more lethal than the G-agents. Of the latter, soman is much more lethal and rapid in action than sarin which, in turn, is about three times more lethal than tabun.

Tabun

Tabun was first discovered in 1936 by Gerhard Schrader, a chemist employed by the German firm I. G. Farben. A colorless liquid with a "fruity smell," it was first produced in industrial quantities in 1942 in Silesia. The Iraqis used it during their war with Iran.

Sarin

Sarin, discovered in Germany in 1938, is a colorless liquid with no smell. Easily volatile, it is mainly taken into the body by inhalation. Sarin has not yet been used in a large amount during warfare. Though it is somewhat more difficult to make than tabun, its preparation is well within the capabilities of a terrorist group.

Soman

Soman, discovered in Germany in 1944, is a colorless liquid with a "fruity smell." So far as is known, it has never been used in warfare. Soman is a fairly volatile substance, taken into the body by inhalation or through the skin.

VX

VX was discovered in 1953 by Ranaji Ghosh, working for ICI, Gerhard Schrader, working for Bayer, and Lars-Erik Tammelin of FOA, the Swedish Institute of Defence Research, all working independently. It too is a colorless liquid with no smell. It is a very persistent substance, like a nonvolatile oil, and will remain on material and the ground for months. The body takes it up through the skin or through inhalation if in gaseous or aerosol form.

How chemical-warfare agents are spread

A chemical-warfare munition converts its payload of chemical agent into an aerosol, a cloud of particles, or a vapor. The types of chemical munitions are similar to those of biological weapons. In an artillery shell, for example, a cylinder of high explosive is

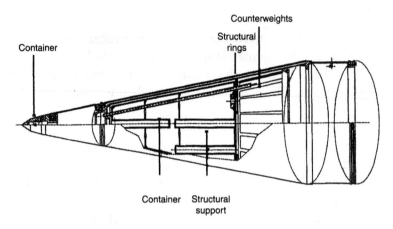

Schematic diagram of chemical weapon warhead (Iraqi special CBW)

aligned along the axis of the shell, which is then filled with the chemical agent. The exploding shell disperses the agent.

A chemical munition is, of course, dangerous in itself. The nerve agent may, for example, leak out of the munition or may be accidentally dropped from a height on to a hard surface. Consequently, chemical weapons called binary weapons have been developed that are safer to handle and easier to store. A binary chemical weapon contains two chemicals. On its own, neither of them is very poisonous, but when they are mixed together they produce a nerve gas. The two chemicals are kept separate until the munition is fired. When it is fired, the chemicals are mixed together so that when the target is hit an aerosol cloud is produced. Before use, one of the chemicals is stored separately.

What chemical-warfare agents do

Blister agents produce large watery blisters on exposed skin that heal slowly and may become infected. Blister agents may also damage the eyes, blood cells, and respiratory tract. By attacking an enzyme, the inhalation of a blood agent prevents the synthesis of molecules used by the body as an energy source, causing vital organs to stop functioning.

Nerve agents (whether dispersed in the form of a gas, an aerosol or a liquid) enter the body through the skin, by inhalation, or by ingestion. They attack the nervous system, so that within minutes of a significant exposure, increasingly severe symptoms appear.

Nerve agents are organophosphorus compounds (as are sheep dip and other insecticides). In the body, they prevent acetyl-cholinesterase, an enzyme essential for the normal functioning of the nervous system, from acting normally. Nerve impulses are transmitted between nerve fibers and various organs and muscles by the compound acetylcholine. When acetylcholine has done its job it is destroyed by acetylcholinesterase, so that the nerve fibers can transmit more impulses. The action of the nerve agent is to inhibit the acetylcholinesterase so that it cannot break down the acetylcholine, which then accumulates, eventually blocking the nerve function.

The initial symptoms will vary according to which agent the individual is exposed to and the amount of the agent absorbed into the body. When an individual is exposed to low amounts of a nerve agent, as a gas or aerosol, the initial symptoms are a running nose, contraction of the pupils of the eyes, blurred vision, uncontrollable crying, headache, slurred speech, nausea, vomiting, hallucinations and reduced mental capabilities, urinary distress, pronounced chest pains, and increased production of saliva. At higher doses, a person will suffer severe coughing and breathing problems, convulsions, deep coma, and finally death. At even higher doses, the symptoms will occur very rapidly and the person will die from suffocation as both the nervous and respiratory systems fail at the same time.

A minute drop of a nerve gas is enough to kill within about twenty minutes

A minute drop of a nerve gas, inhaled or absorbed through the skin or eyes, is enough to kill within about twenty minutes.

The reality of chemical attack: eyewitness accounts

The effects of a chemical weapon attack are described by John Simpson, the BBC's Foreign Affairs Editor, who visited Halabja six days after the Iraqis attacked the town with chemical weapons, reportedly including mustard gas, nerve agents and cyanide. Iranians had captured the town from the Iraqis on March 15, 1988. The next day, the Iraqis attacked with chemical weapons, delivered by combat aircraft and artillery. The attack lasted for several hours.

John Simpson described what he saw as he walked from the outskirts of Halabja towards the town center:

The first thing I noticed was the stench, then the bodies of sheep and cows starting to bloat and swell. Out in the fields there were occasional shapes under blankets, shepherds whose bodies had not yet been taken away. As we approached the streets of the town we found that many—though not all—of the bodies had been left for us to see. By now they had all been dead for five days. A street we turned into was full of them. No one had touched them since they had run out of their homes, hoping to get away from the houses and streets where the gas would collect. Some of them had shrapnel wounds but it wasn't the shrapnel that killed them. There was a strange waxiness to their faces, an absence of fear or pain. Many of them lay with their eyes open and, even after five days out in the open, they seemed more human than the usual bundles of dead flesh in old clothes which you find after massacres.

> *The first thing I noticed was the stench, then the bodies of sheep and cows starting to bloat and swell*

A Kurdish eyewitness recounted:

I heard the planes coming over, and then I saw the bombs starting to fall. White smoke came out of some of them. I was near the concrete shelter, so I went down into it and waited. I was too frightened to come out for a long time. When I did, it seemed as though the whole town was lying dead. People I knew, just lying in the streets, or lying inside their houses. It was a dead place, this town.

Simpson looked at a group of bodies lying near him,

a young woman, perhaps twenty, in a magenta and orange dress, holding a baby in her arms. The mother could have been sleeping, but the baby's eyes were white and dead. Its clothes were continually fluttering in the slight wind.

We wandered around the houses. Most of them still had people inside. I went into one where a rocket had come through the roof. There was a sound of buzzing: flies were at work on the food the family had been eating when the attack began. There were six of them around the table. A child had rolled out of his chair and lay on the floor, face down. A man and a woman were hunched down in their seats: I couldn't see their faces. An older man, the grandfather, lay with the side of his face on the table, his hand to his mouth, his jaw still clamped on a piece of flat bread which he had been in the act of biting when the rocket came through the roof and filled the room with poison gas.

I sniffed the air: mustard gas smells of sewage; nerve gas has a much more pleasant smell, like chocolate according to some people and new-mown hay according to others; cyanide gas supposedly smells like almonds, though if you take a single breath of it you are likely to die. According to the Iranian doctor who accompanied us it was cyanide that killed the old man eating his bread and the mother with the child in her arms. All over in a second or two, he said. The others were really bad. Nerve gas strangles you from the inside and takes a long time to do it. The contorted bodies I saw later on the edge of town must have died that way. Mustard gas kills only 3 percent of people on average, but condemns a good 50 percent of the others to an ugly half-life of chest and throat pain, of huge blisters which can erupt 10 or 20 years later, of serious damage to the eyes and the nasal passages.

We went on. More bodies, then an entire truck full, four or five deep: dozens of women in bright-colored clothes, dozens of old men and children. The stench was more than I could stand and the sight of those calm, gray-white faces was beginning to haunt me.

Halabja victims

The British Ministry of Defence announced in 1999 that "the potential threat from biological and chemical weapons is now greater than that from nuclear weapons." The threat is greater because more countries have biological and chemical weapons, which in turn is because these are more easily and cheaply produced than nuclear weapons.

On June 8, 2001, George W. Bush was more specific, nominating biological warfare as the greatest threat faced by the United States and its allies. Biological and chemical weapons are undoubtedly frightening weapons, designed for deliberate mass injury and death. Most people, however, fear nuclear weapons more than they fear biological or chemical ones. This is why the nuclear-weapon powers regard their weapons as a far greater deterrent than biological and chemical ones would be. In fact, France, Russia, the UK and the USA do not even have biological or chemical weapons in their arsenals.

Biological and chemical weapons generate severe psychological effects in people. To be struck down by a dread disease or poisoned by some terrifying chemical are nightmare scenarios. Descriptions of what happens to humans attacked by biological and chemical weapons show how well founded fears of these weapons are.

Biological and chemical weapons attack humans mainly through the respiratory tract—by breathing them in—although some chemical-warfare agents attack through the skin. In addition, biological weapons can be ingested in food or water. Biological and chemical munitions operate by dispersing agents into the air. Anyone downwind is then attacked indiscriminately. A typical chemical-warfare attack would involve tons of a chemical agent; a typical biological-warfare attack would involve mere kilograms of a biological agent.

Chemical terrorism

Of the three types of weapons of mass destruction—chemical, biological, or nuclear—chemical ones are the most accessible to terrorists. The chemical weapon of choice for terrorists would probably be a nerve agent. The details of the chemical preparation of nerve agents are described in the open literature, including scientific journals, chemistry textbooks and the World Wide Web. The chemicals required to manufacture nerve agents are readily available and no specialized chemical equipment is needed for the preparation; a competent chemist would have no difficulty in making tabun, sarin, or soman, all with ingredients legally obtainable. Tabun is the easiest nerve agent to prepare and, therefore, likely to be of particular interest to terrorists. The chemicals needed to prepare tabun—dimethylamine, sodium cyanide, and phosphoryl chloride—can be obtained on the open market.

Of the three types of weapons of mass destruction— chemical, biological, or nuclear— chemical ones are the most accessible to terrorists

Having made a nerve agent, terrorists would need to disperse it. They could, for example, make or acquire a device to produce an aerosol so that the nerve agent is released as a cloud of droplets. The device, perhaps set off by remote control, could be placed to produce an aerosol in, say, a city's underground train system. If this were done effectively, a very large number of people would be killed. An aerosol could also be released into the ventilation system of, for example, a large office block.

The AUM Shinri Kyo (Supreme Truth) Cult demonstrated that a terrorist group could manufacture and disperse a nerve agent. On March 20, 1995, shortly after 8 a.m., the group released sarin on five different underground trains converging on the Kasumigaseki station in Tokyo. The attack killed twelve people and injured more than five and a half thousand, 500 of whom had to be hospitalized. The Tokyo attack followed another, little-publicized, sarin gas attack in June 1994 in a small town north of Tokyo, that killed seven and injured about 200 people.

PART II

Weapons of mass destruction and the state

4 What does it take to make a WMD?

The facilities and resources needed to produce WMDs, and to produce and deploy the munitions to deliver them, vary according to the type of WMD; a program to produce and deploy a nuclear-weapon force is a much larger undertaking than one to produce a biological- or chemical-weapon one. The acquisition of a nuclear force, large enough to be strategically significant within the region, requires the investment of a huge sum of money and the employment of a very large group of specialists. Nevertheless, a number of developing countries—Israel, India, and Pakistan—have made the investment and deployed a nuclear force; North Korea has a nuclear-weapon program, Iran may be planning one, and Iraq was developing nuclear weapons before the 2003 war.

all but the least developed countries could afford the resources to deploy chemical weapons

A program to deploy a chemical-weapon force requires the investment of the least amount of resources and is the least demanding of the three; all but the least developed countries could afford the resources to deploy chemical weapons.

What is a nuclear-weapon program?

A country that does not have nuclear weapons but does have advanced nuclear technology and security concerns in its region is often called a "latent nuclear-weapon state" or a "virtual nuclear state." Japan is an example of such a country. A country without such advanced nuclear technology may decide for reasons of security or prestige to acquire nuclear weapons. Iraq, Iran, and North Korea are suspected of doing so. What materials, facilities, and personnel would such countries need to become actual nuclear-weapon powers?

A country with a civil nuclear program will have little difficulty in designing, developing and fabricating nuclear weapons. If it is, for example, using nuclear-power reactors to generate electricity, it will already have the skilled physicists, chemists, technologists, engineers, and technicians and the capability to produce plutonium suitable for use in nuclear weapons. The "peaceful atom" and the "military atom" are intimately linked— "Siamese twins" in the words of Nobel Prize winning Swedish physicist, Hannes Alven.

Thirty countries are operating 438 nuclear-power reactors for the generation of electricity today. They include developing countries, such as Argentina, Brazil, India, Mexico, Pakistan, and South Africa. Countries also operate reactors, called research and test reactors, to produce radioisotopes for medical, industrial, and agricultural use and for training physicists and engineers. (Radioisotopes are used in medicine to diagnose and treat diseases; in industry to radiograph large structures; and in agriculture to kill pests and sterilize male insects to reduce their numbers.) A country with this type of reactor also has a group of skilled people who could be diverted to a nuclear-weapon program.

Military scientists in any industrialized country will, it can be assumed, be collecting information on nuclear weapon design—by searching the scientific and technical literature, attending relevant meetings and conferences, engaging in espionage, and so on. Perhaps only a small number of people working in a military research establishment will be involved in this preliminary stage. A team may be set up actually to design nuclear fission weapons, boosted weapons, and perhaps thermonuclear weapons. Computer simulations of weapon design and the effects of a nuclear explosion may well be performed. The computer codes needed for these activities are available commercially. This preliminary work may be said to be "for defensive purposes." There is unlikely to be a specific political decision to undertake this work.

The move to an active nuclear-weapon program will almost certainly require a political decision, which may be taken by the country's political leader, perhaps with discussion with a small number of colleagues, but need not involve the whole cabinet. In the United Kingdom, for example, the decision to acquire nuclear weapons was taken by Prime Minister Clement Attlee, Foreign Secretary Ernest Bevin and the Minister of Supply, whose department was responsible for the program. In France, it was taken by President de Gaulle.

In the United Kingdom, the decision to acquire nuclear weapons was taken by Prime Minister Clement Attlee

The first step of the program will be to acquire nuclear material—plutonium and highly enriched uranium. Industrialized countries will want to produce these indigenously so that they are not dependent on others for them. Indigenous production will require the acquisition of special materials—such as a specially strong steel (called maraging steel), and special tubing, for the construction of gas centrifuges to produce highly enriched uranium from natural uranium—or materials to construct a nuclear reactor for the production of plutonium from the uranium. If the country has a civil nuclear program, it will already have a team of nuclear physicists and engineers, some of whom can be used in a nuclear-weapon program.

At this stage there will be experiments to develop implosion technology to compress a sphere of fissile material into a super-critical mass, including the development of pure conventional high explosives. Experiments will be carried out, probably in an establishment run by the defense ministry, to implode spheres made from non-fissile material, such as natural or depleted uranium.

When it has been demonstrated that fissile materials can be successfully produced, a political decision will be needed to go ahead and produce and deploy nuclear weapons. The initial size of the nuclear force will be also discussed. At this stage decisions will be required about delivery systems—combat aircraft or, more likely, surface-to-surface missiles. New systems will be developed or existing ones modified. Groups will be established in the Ministry of Defense to monitor and develop the various military activities.

The military will need to integrate nuclear weapons into tactics and strategy, which will mean evolving these processes for nuclear use and developing and setting up an effective command, control and communications system. War games and military maneuvers will be undertaken and textbooks will be prepared for use in military colleges. Aircrew will practice the air delivery of nuclear weapons. The pilot will be trained to maneuver the aircraft after dropping the nuclear weapon to avoid being damaged by the effects of the explosion.

A nuclear-weapon program involves decisions and activities by the country's military scientists and engineers, the political leaders, the military leaders, defense bureaucrats, industry, and academics. A separate military-political-industrial-bureaucratic-academic complex will evolve devoted to the production and deployment of nuclear weapons and the development of tactics and a strategy for their use.

What do you need to make a nuclear weapon?

Both civil and military nuclear programs depend on uranium. Uranium was discovered in 1789 by the German chemist H. M. Klaproth but was used for only minor purposes—in, for example, chemistry and metallurgical research—until the Second World

War when large quantities began to be used in nuclear-weapon programs. Since the 1950s large amounts have been and are being used to fuel nuclear-power reactors.

Uranium is a very widely distributed element found, as an oxide, in a large variety of minerals and in seawater. Most of the uranium is dispersed through the rocks of the Earth's crust; only a small fraction is found in concentrated ores. Deposits that can be mined economically occur in sandstones, shales, granites, phosphates, lignites, and quartz-pebble conglomerates and veins.

Uranium is mined in open-pit and underground mines. Uranium mines are operated in about twenty-four countries. The biggest ones are in Australia, Canada, Namibia, Niger, Russia, South Africa, and the USA. In addition, China, the Czech Republic, France, Kazakhstan, Tajikistan, and Uzbekistan mine significant amounts. There is a world glut of uranium, so any country intent on doing so could readily get its hands on supplies.

A country running a clandestine nuclear-weapon program will, if it can, mine its own uranium. Israel, for example, mines uranium in the Negev desert with phosphate deposits. India mines uranium at Jaduguda and Pakistan mines it at Dera Ghazi Khan. Iran has recently opened a uranium mine about 200 kilometers from the city of Yazd.

Uranium mining is a hazardous activity. There are not only the usual dangers of mining, but also radioactive decay products which accompany the uranium to contend with. Both uranium-235 and uranium-238 are radioactive and both have a family of daughter products. Uranium-238, for example, has fourteen radioactive daughter products, one of which is the gas radon. If a uranium miner breathes a radioactive dust particle or radon he or she could get lung cancer. People living in badly ventilated houses built on granite or containing granite are also exposed to radon and run a risk of a similar kind.

If a uranium miner breathes a radioactive dust particle or radon he or she could get lung cancer

Once mined, the uranium ore, often still in rock, is taken to a

The world's top uranium-producing countries

uranium mill. The amount of uranium in typical ore is only one or two parts per thousand. The uranium mills are, therefore, usually close to the mine. The ore is crushed, mixed with water, and ground into fine particles. The mixture is put through a chemical procedure to purify it. This produces a uranium oxide, U_3O_8, a yellow compound called yellow cake, which is sold in this form for about 22 US dollars per kilogram.

Every thousand atoms of naturally occurring uranium contains only seven atoms of uranium-235; the other 993 are atoms of uranium-238. This concentration of uranium-235 is too low to produce the supercritical mass needed to generate a fission chain reaction in a nuclear weapon. Therefore, the concentration of uranium-235 in uranium is increased in a process called enrichment.

The extent of the enrichment depends on the purpose for which the uranium is required. Some military reactors used to produce plutonium for use in nuclear weapons are fueled with natural uranium and use no enriched uranium. Commercial nuclear-power reactors use uranium enriched to about 4 percent in uranium-235. For use in a fission nuclear weapon, uranium is enriched to more than 90 percent.

Uranium-235 and uranium-238 are chemically identical and so it is necessary to use a physical method to separate and enrich them. The difference between the two isotopes is that the nucleus of a uranium-238 atom contains three more neutrons than the nucleus of a uranium-235 atom, giving a minute difference in the weight of the atoms. There are two main methods of using this difference to separate the isotopes, using a gaseous diffusion method or gas centrifuges. Both methods use a uranium gas, uranium hexafluoride, which is a solid at room temperature. It is converted into a gas by heating it to a temperature of about 64 degrees Celsius. Pure uranium hexafluoride is obtained by converting yellow cake in a chemical plant, called a conversion plant.

Gaseous diffusion relies on the fact that in a gaseous mixture of the two isotopes, the molecules of uranium-235, the lighter one, will diffuse more rapidly through a porous barrier than the molecules of uranium-238, the heavier one. Uranium hexafluoride is extremely corrosive and reactive, so much so that special materials, such as nickel and aluminium alloys, have to be used for the construction of the pipes and pumps used in a diffusion plant, and the entire installation must be kept free of grease and oil so as not to produce undesirable chemical reactions with the hexafluoride.

The proportion of uranium-235 in the gaseous mixture is increased by only a small fraction in each diffusion stage. Numerous stages are required to give a significant enrichment. A commercial diffusion plant uses an enormous amount of electrical power, usually requiring the construction of an independent power station. A large diffusion plant is operating in each of China, France, and Russia; two are operating in the USA. These plants were all originally built for military purposes, to produce highly enriched uranium for use in nuclear weapons. They are now mainly used to produce enriched uranium to fuel nuclear-power reactors. Argentina operates a small pilot diffusion plant but has not yet taken a decision to construct a commercial plant.

The gas centrifuge method of enriching uranium also relies on the minute difference in mass between uranium-235 and uranium-238 atoms. It uses a rapidly spinning centrifuge to separate the isotopes. The centrifuge consists of a cylindrical drum that rotates at very high speeds. The heavier uranium-238 concentrates at the outer radius of the drum and is made to flow in one direction, while the uranium-235 is enriched near the central axis of the drum and is made to flow the opposite way. The enriched uranium-235 is collected through an exit orifice.

Although the separation of the uranium isotopes is much greater per stage in a centrifuge plant than in a diffusion plant, it is still very small. A centrifuge plant, therefore, contains a great number of centrifuges in a cascade to achieve a useful output of enriched uranium. The slightly enriched uranium-235 from the first centrifuge in the cascade is fed into the input nozzle of the next centrifuge, the slightly more enriched uranium-235 from the second centrifuge is fed into the third, and so on.

The requirements for the materials used in the construction of

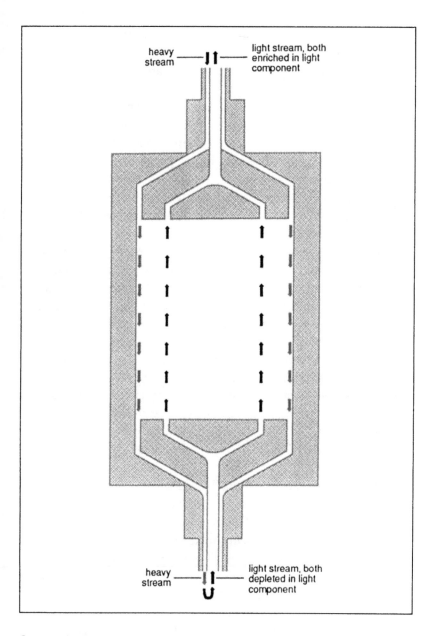

Gas centrifuge

gas centrifuges are very demanding. The outer casing of the drum and the rotor bearings, in particular, must be made from a material with a high tensile strength, the most suitable of which is carbon fiber.

Gas centrifuge plants are cheaper to run than diffusion plants. The amount of electricity required to operate a centrifuge plant is typically about one tenth of that required for a diffusion plant. Another advantage with a centrifuge plant is that it can be built up in stages, step by step, as demand for enriched uranium increases. Large gas centrifuge plants for uranium enrichment operate in China, Germany, Japan, the Netherlands, Pakistan, Russia, and the UK.

Electromagnetic separation has been used to enrich uranium on a significant scale, in machines called calutrons. It was used in the Manhattan project in the Second World War to produce the highly enriched uranium for the Hiroshima bomb. The USA stopped using calutrons in 1946 because they were very expensive to operate. More recently, Iraq experimented with the technique but soon abandoned it in favor of gas centrifuges.

In a calutron, atoms of uranium are ionized—that is, one or more electrons in the atom are removed—and injected into a magnetic field. The particles bend as they travel in the magnetic field with the lighter particles, the uranium-235 particles, bending more than the heavier uranium-238 particles. Separation can thus be achieved.

The South Africans are using the helicon or jet nozzle method of separating uranium isotopes at a plant at Valindaba. The process, developed at Karlsruhe in Germany, is an aerodynamic one, using pressure diffusion in a gaseous mixture of uranium hexafluoride and a light gas, such as helium or hydrogen, flowing at high speed through a nozzle along sharply curved walls. The heavier molecules are less deflected in the stream with the largest curvature, allowing separation to take place.

For the enrichment of uranium to the extent needed to produce nuclear weapons, normally uranium containing more than 90 percent of uranium-235 is used. To produce 1 kilogram of this uranium requires about 180 kilograms of natural uranium. Each nuclear weapon typically contains about 15 kilograms of highly enriched uranium, requiring the mining of about 1,500 tons of uranium ore.

Plutonium

If a country decides to produce plutonium to use as the fissile material in its nuclear weapons, it will need to construct two key facilities, a plutonium-production reactor and a chemical plant to separate, or reprocess, the plutonium from other materials in the fuel elements when they are removed from the reactor.

All reactors produce plutonium; military plutonium-production reactors do so very efficiently. Unlike nuclear-power reactors, they produce no usable power or energy.

Plutonium results when uranium-238 absorbs some of the neutrons produced in the fission process, to become the isotope uranium-239. Uranium-239 is radioactive and decays to plutonium-239. This plutonium isotope can be used as the fissile material in nuclear weapons.

The uranium fuel elements are removed from a plutonium-production reactor after a short time, typically three months or so. At this time, the plutonium is of the type best suited for use in nuclear weapons. If the fuel elements are left much longer the plutonium-239 itself absorbs neutrons, producing plutonium-240 and plutonium-241. These other isotopes contaminate the plutonium-239 and the fewer there are of them the better.

When the fuel elements are removed from the plutonium-production reactor they contain, in addition to the plutonium-239 and a small amount of plutonium-240 and -241, unused uranium and fission products. The reprocessing plant chemically separates the plutonium from the uranium and the fission products. The method used in reprocessing plants is

A reprocessing plant is an essential facility for the production of plutonium for nuclear weapons

generally the PUREX process in which tributyl phosphate and kerosene are used to separate the fission products from the uranium and plutonium.

A reprocessing plant is an essential facility for the production of plutonium for nuclear weapons. The existence of one suggests that a country intends to use plutonium to produce nuclear weapons. All countries with significant reprocessing facilities are, therefore, actual or potential nuclear-weapon powers.

A nuclear weapon normally contains about 4 kilograms of plutonium. A country wanting to produce, say, three nuclear weapons a year will need a reprocessing facility able to separate about 12 kilograms of plutonium a year. A facility of this capacity is small enough in physical size to be easily hidden, and so is a small plutonium-production reactor able to produce 12 kilograms of plutonium a year. Either facility could be constructed in a moderately sized two-story building. A country which has decided to use plutonium rather than highly enriched uranium could, therefore, do so clandestinely. It would be more difficult to conceal a gas centrifuge plant able to produce enough highly enriched uranium to produce three nuclear weapons a year, as satellite photographs would be able to spot the construction and operation of such a plant.

The fuel elements removed from a plutonium-production reactor are very radioactive and have to be handled with remote equipment. Also, parts of the reprocessing plant have to be heavily shielded to prevent the workers becoming exposed to too much radiation.

The production of the components for nuclear weapons

The plutonium will leave the reprocessing facility as plutonium dioxide. This will be converted into plutonium metal. The metal is then rolled, formed and heat treated to produce small plutonium ingots, each weighing less than a kilogram. These operations require special equipment such as furnaces and lathes. The ingots are shaped in a foundry into the solid or hollow spherical form of

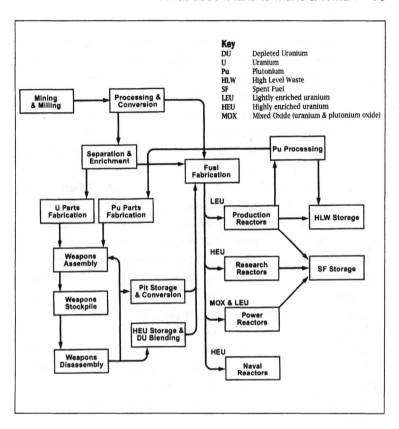

a weight, of about 5 or 6 kilograms, suitable for use in the core of nuclear weapons.

These operations will be performed in an atmosphere of an inert gas. Because of its high toxicity, no plutonium will be allowed to escape into the human environment. Care in handling toxic materials is required during the production of other components for a nuclear weapon, such as the manufacture of the beryllium shell used as a neutron reflector.

The chemically pure conventional high explosives, normally HMX, used to produce the symmetrical shock wave to compress the plutonium (or highly enriched uranium) in the core of the weapon will be fabricated in a special establishment with facilities to test and perfect the implosion system.

If highly enriched uranium is used as the fissile material in nuclear weapons, the enriched hexafluoride will be first converted into uranium oxide and the oxide will then be converted into uranium metal, procedures that are straightforward chemistry. Ingots of highly enriched uranium metal will be produced using processes similar to those used to produce plutonium ingots.

If the "gun" design is used in the nuclear weapon, two masses of highly enriched uranium, each less than a critical mass but together making a supercritical mass, are produced. When the two masses are assembled together, a nuclear explosion will occur.

If the implosion technique is used, the highly enriched uranium is machined into components that can be put together to produce either a solid or a hollow sphere. The total mass of highly enriched uranium will be less than the critical mass. This core is surrounded by conventional high explosive to compress the highly enriched uranium to produce a supercritical mass and a nuclear explosion.

The critical mass of a bare sphere of highly enriched uranium is much greater than the critical mass of a bare sphere of plutonium-239; the former is about 52 kilograms, the latter is about 11 kilograms. These critical masses can be reduced by more than half by surrounding the core of fissile material with a thick neutron reflector made from, for example, beryllium.

Because of the relatively small amount of plutonium-239 needed in a nuclear weapon, it is normally the material of choice in nuclear-weapon programs. Of the current nuclear-weapon powers, only Pakistan uses highly enriched uranium in its weapons. South Africa also used highly enriched uranium to produce six nuclear weapons, based on the gun technique. The weapons were built and deployed in the late 1970s. Following a decision in 1989, taken by former President F. W. de

plutonium-239 is normally the material of choice in nuclear-weapon programs

Klerk, the weapons, all based on the gun technique, were dismantled, making South Africa the only country to deploy nuclear weapons and then dismantle them.

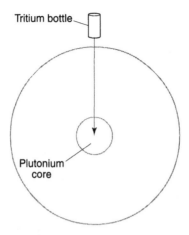

The Tritium bottle is an external component, for ease of change

If a country decides to boost its nuclear weapons to increase their explosive power, it will also need a facility to produce tritium. Tritium is the radioactive isotope of hydrogen; each tritium atom contains a proton and two neutrons in its nucleus. It can be used to produce nuclear fusion (see diagram). Tritium is produced in a suitable nuclear reactor. Its production is another reason why a country planning nuclear-weapon force will acquire a nuclear reactor. Israel, for example, uses its reactor at Dimona to produce both plutonium and tritium for use in its nuclear weapons.

Nuclear testing

A nuclear weapon using just nuclear fission to produce the energy for a nuclear explosion does not need testing. The design is so straightforward and well tried that the scientists and technologists who produce the weapons can be confident that they will work without testing. Countries having a very competent nuclear community, like Israel, will probably not need to test boosted nuclear weapons but would need to test thermonuclear ones.

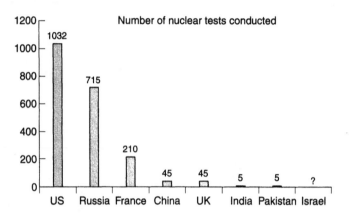

Nuclear tests by country

Seven nuclear-weapon powers (China, France, India, Pakistan, Russia/the Soviet Union, the United Kingdom and the United States) are known to have conducted a total of at least 2,052 nuclear tests since the first nuclear explosion in 1945. The USA carried out its first test of a nuclear fission weapon on 16 July 1945, the former Soviet Union did so on September 23, 1949, the UK on October 2, 1952, France on October 13, 1960, China on October 16, 1964, India on May 11, 1998, and Pakistan on May 28, 1998. The dates on which countries that have thermonuclear weapons made their first full-scale thermonuclear test explosion are: USA, November 1, 1952; the former Soviet Union, August 21, 1953; the UK, May 15, 1957; France, September 24, 1968; and China, June 17, 1967.

The USA carried out a total of 1,032 tests; the former Soviet Union made 715; France 210; China 45; the UK 45; India 5; and Pakistan 5. Israel, the eighth known nuclear-weapon power, has not, so far as is publicly known, tested a nuclear weapon although an explosion high in the atmosphere on September 22, 1979, off the eastern coast of South Africa, is widely believed to have been a clandestine Israeli nuclear test. There have been no known nuclear tests since the end of 1998.

American nuclear tests were performed at sites in Nevada and in the Pacific; the former Soviet Union had test sites in Kazakhstan and Novaya Zemlya; the British had test sites in Australia, Christmas Island, and Nevada; the French in Algeria and Polynesia; the Chinese at Lop Nor, India at Pokhran in the Thar desert; and Pakistan at Chagai Hills.

Until 1981, many nuclear tests were carried out in the atmosphere; since 1981 they have all been performed underground. The total explosive yield of all the nuclear tests conducted so far is equivalent to that of roughly 510 megatonnes of TNT, or the explosion of about 40,000 Hiroshima bombs. The atmospheric tests are responsible for much radioactive contamination of the human environment, fifty times more than that released by the 1986 Chernobyl nuclear accident. It has been estimated that exposure to the radiation from the radioactivity produced by atmospheric nuclear tests will eventually cause the death of about 1.5 million people.

exposure to the radiation from the radioactivity produced by atmospheric nuclear tests will eventually cause the death of about 1.5 million people

Nuclear tests are performed to develop knowledge about nuclear fission weapons and the effects of nuclear explosions, and to conduct initial research into thermonuclear explosions. Operational nuclear and thermonuclear warheads were tested, including a variety of actual nuclear munitions such as artillery shells and aircraft bombs. Troops were exposed to nuclear explosions during some of the earlier tests to analyse the effects of the explosions on the battlefield.

Underground nuclear tests are normally conducted in an excavated chamber, deep enough to contain the radioactivity produced by the nuclear explosion. Scientific instruments and equipment to measure the effects of the nuclear explosion are placed in tunnels running off the chamber. In some cases, the blast from the explosion has broken through the surface and radioactivity has escaped into the atmosphere, a process called venting.

The making of chemical weapons

Just as biological-warfare agents are made in plants very similar in design to civilian plants used to produce high-grade pharmaceutical products, so nerve agents for use in chemical weapons can be made in plants very similar to industrial chemical plants used to manufacture herbicides. Both nerve agents and herbicides are organophosphorous compounds.

The military chemical plants will employ chemists, industrial chemists, physical chemists, and technicians. The plants will include measures to contain the toxic agents to prevent the workers becoming exposed to them, including the use of barriers to separate workers from the organophosphorous compounds and efficient methods of ventilation.

Biological and chemical munitions

The ordnance for the delivery of biological (and chemical) weapons, including artillery shells, mortar rounds, rockets, aircraft bombs, and missile warheads, is produced in special factories. Great care must be taken to prevent any leakage of biological and chemical agents into the human environment. The workers in the factories are protected by effective containment of the agents, the use of physical barriers to separate the workers from the agents, and effective ventilation.

Biological and chemical munitions normally disseminate liquid agents as aerosols. Both types of agents are most lethal if the aerosols produce drops of liquid that are of a consistent and appropriate size. The droplets should be about a micron (a millionth of a meter) in diameter, a size that makes it possible to breathe them deep into the lung. Larger droplets are filtered out by the nose and do not get into the lung. The munitions must, therefore, be carefully designed. Steps must be taken to enable the munitions to deliver the agent to the target with minimum degradation.

5 Case studies: Iraq and North Korea

Iraq is an example of a developing country with biological-, chemical- and nuclear-weapon programs. It has successfully developed and deployed a variety of biological and chemical weapons and made good progress in its program to produce indigenously fissile materials for nuclear weapons. Iraq has not succeeded in producing nuclear weapons. North Korea probably has.

Iraq is by no means the only country in the Middle East with WMD programs. Israel has nuclear and probably also biological and chemical weapons. Other countries in the region which possess, are developing or are capable of producing WMDs are Egypt (biological and chemical), Iran (biological, chemical and nuclear), Libya (biological and chemical), Pakistan (biological, chemical and nuclear), Sudan (chemical), and Syria (chemical).

Iraq's nuclear capability

Attempts to produce plutonium

Iraq began its nuclear research activities in 1956 by setting up the Iraqi Atomic Energy Commission. In 1967, a research reactor,

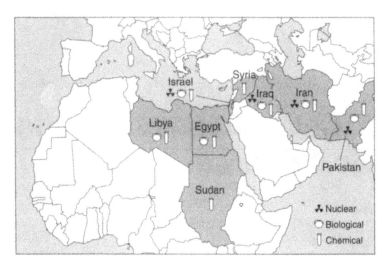

Nuclear, chemical, and biological powers in the Middle East

supplied by the former Soviet Union, began operating at the Nuclear Research Centre at Tuwaitha, near Baghdad. Iraq significantly expanded its nuclear program in the mid-1970s, making an agreement with France for the supply of a larger research reactor.

This reactor, called Tammuz, that could have produced about 7 kilograms of plutonium a year, was bombed and destroyed by the Israelis in a sudden air attack on July 7, 1981 before the fuel was loaded into it. The Israelis, who by then had deployed a significant nuclear-weapon force of its own, also with French assistance, did not want a nuclear-armed Arab country as a neighbor.

The Israeli raid put a stop to Iraqi ambitions to acquire significant amounts of plutonium although Iraq acquired another small research reactor from the former Soviet Union; it began operating at Tuwaitha in 1987, but was not capable of producing more than small amounts of plutonium.

The author visited the Nuclear Research Centre at Tuwaitha in the late 1970s and talked to senior nuclear scientists there. They were highly competent professionals. Typically, an Iraqi nuclear physicist had spent time at one of the world's best nuclear research centers, such as the CERN center in Switzerland. At the time, there were only a few such highly educated nuclear scientists with side experience at Tuwaitha. But only a handful of such people are required to set up a nuclear-weapon program.

In the early 1980s, Iraq turned its attention to the clandestine production of highly enriched uranium for use in nuclear weapons rather than plutonium. Our knowledge of this program is based mainly on the inspections made by International Atomic Energy Agency inspectors (the IAEA Action Team). These inspections began after the Gulf War in 1991 and went on until 1998 when the inspectors were expelled by Iraq. Since 1998, information about Iraq's nuclear activities has come mainly from Iraqi defectors, often not a very reliable source.

Since 1998, information about Iraq's nuclear activities has come mainly from Iraqi defectors, often not a very reliable source

The IAEA team discovered that by 1991 Iraq had experimented with no fewer than five methods for the enrichment of uranium—gaseous diffusion, laser enrichment, chemical enrichment, gas centrifuges, and calutrons. It is believed to have invested much more than a billion US dollars, over nearly two decades, in investigating these five techniques.

It eventually decided to concentrate on two methods: large calutrons and gas centrifuges. Production facilities for the former were established at Al Sharqat and Al Tarmiya. Just before the first Gulf War began Iraq decided that gas centrifuges were a more attractive way of enriching uranium and gave up calutrons. Although calutrons are technologically relatively simple, and within Iraq's capabilities to construct and operate, they are also very expensive and inefficient.

But the indigenous establishment of a clandestine gas centrifuge plant presented Iraq with serious difficulties. Gas centrifuge technology is much more technically demanding than calutrons and requires specialized materials and equipment, much of which had to be acquired and imported from abroad. After 1988, valuable assistance was obtained from foreign firms, particularly German ones, and some German centrifuge specialists were recruited by the Iraqis. The Iraqis succeeded in getting a surprising amount of foreign assistance from European gas-centrifuge experts and materials from European firms, particularly the German firm H&H Metalform. Swiss firms also helped.

Nevertheless, the Iraqis did even less well with their gas centrifuges than with calutrons. Gas centrifuges contain rotors—cylinders rotating at very high speeds—made from special materials, particularly maraging steel or carbon fiber. Based on its experience before 1991, Iraq knew how to make carbon fiber but it was not able to make its own centrifuge rotors. Its stock of carbon fiber was handed over to UNSCOM inspectors.

Some maraging steel was imported from Germany and other countries. A typical centrifuge for Iraq would probably consist of two rotor tubes, made from maraging steel or carbon fiber, connected by maraging-steel bellows (flexible joints). Iraq would probably have preferred rotors made from carbon fiber rather than maraging steel because the former can spin faster than the latter, which is limited to a speed of about 450 meters per second.

Constructing the bellows is not an easy task. The bearings are also difficult to get right. By the end of the Gulf War in 1991, the Iraqis had produced no significant amount of highly enriched uranium by gas centrifuges or any other method. United Nations weapons inspectors comprehensively destroyed Iraq's embryonic indigenous uranium-enrichment capability.

By the end of the Gulf War in 1991, the Iraqis had produced no significant amount of highly enriched uranium

The Iraqis were more successful in nuclear weapon design than in highly enriched uranium production. The IAEA also discovered that Iraq had evolved an effective nuclear-weapon design, based on the implosion method, and had assembled the non-nuclear components of a fission weapon of the type that destroyed Nagasaki in 1945.

The IAEA found that Iraq had experimented with high explosives lenses to produce implosive symmetrical shock waves and had developed an electronic firing system for a nuclear weapon using thirty-two detonators arranged symmetrically in the high explosive. The system was tested at the Al Qa Qa facility. The Iraqis had imported a large amount of HMX and RDX explosives and were producing their own RDX.

The critical mass of a bare sphere of weapon-grade enriched uranium is about 55 kilograms. Assuming that Iraq uses its implosion-type design with a sphere of highly enriched uranium and surrounds the highly enriched uranium sphere with a thick shell of a neutron reflector like beryllium, it could cut the critical mass to about 20 kilograms.

Bearing in mind that Iraq cannot test a nuclear weapon, and would therefore use more than an absolute minimum of highly enriched uranium in its first nuclear weapons, it can be assumed that it would need at least 25 kilograms of highly enriched uranium for each weapon of the implosion type. A strategically significant nuclear force for Iraq would consist of at least six nuclear weapons, requiring 200 or so kilograms of highly enriched uranium, allowing for some wastage.

The capacity of a gas centrifuge is measured in separative work units (SWU). A reasonable estimate is that each centrifuge of the type that Iraq is likely to produce would have a capacity of about 2.5 SWU per year. In 1991, Iraq was testing two prototype centrifuges. In one test, a carbon-fiber rotor, provided by the German firm Schaab, was spun at up to 60,000 rpm. The enrichment capacity during the best test run reached 1.9 SWU per year. IAEA inspectors estimated that an output of 2.7 SWU per year could have been achieved, but this would have required much more development work.

At the end of 1990, a factory for the production of centrifuges was being built at Al-Furat and a gas centrifuge plant was planned at Rashdiya. The Rashdiya plant was designed to produce about

10 kilograms of highly enriched uranium a year. This would have been a relatively small plant.

It would take a facility containing 3,000 centrifuges to produce 7,500 SWU per year or about 40 kilograms of highly enriched uranium. It would take this facility at least five years to produce enough highly enriched uranium for the nuclear force of six nuclear weapons.

Assuming that about 60 percent of the centrifuges have to be rejected as substandard, a reasonable assumption, Iraq would need to produce about 5,000 centrifuges for this facility. Moreover, gas centrifuges break down frequently because of the mechanical stresses they are under. A steady supply of replacement machines must, therefore, be produced.

A facility operating a cascade of 3,000 centrifuges would use as much energy, electrical power, as a largish city—approximately 200 kilowatt-hours per SWU or roughly 1,000 kilowatt-hours per gram of highly enriched uranium. In other words, it would be impossible to operate such a facility clandestinely. The Iraqis would have needed constant input of foreign expertise and assistance to run such a plant. Building and operating effectively a gas centrifuge facility of a useful size is not a trivial task—it is an industrial undertaking. It would probably have taken Iraq at least five or six years to build such a facility and begin producing significant amounts of highly enriched uranium. These are somewhat rough and ready estimates, but they clearly make the point that Iraq was some considerable way off being a nuclear power.

Iraq's biological capability

Iraq's biological-weapon program became significant in the 1980s after the outbreak of the Iran–Iraq war. A research and development facility was operated at Salman Pak, near Baghdad, and a production plant, at Al Hakam, began producing biological-warfare agents in 1988. Biological munitions were filled with agents at a facility at Al Muthanna.

After mid-1990, biological-warfare agents were also produced at the Al Dawrah Foot and Mouth Vaccine Institute and at the Al Fudaliyah Agriculture and Water Research Centre. By 1991, Iraq was producing aflatoxin, anthrax, botulinum toxin and ricin toxin. It was also conducting research on a number of other agents, including viruses; and, according to some commentators, on genetic engineering.

Iraq's delivery systems for biological weapons were Al Hussein missile warheads, R-400 aircraft bombs and probably drop-tank spray devices delivered by Mirage F-1 bombers. These munitions were loaded with anthrax, botulinum and aflatoxin. Reportedly, about twenty-five missile warheads, about 160 R-400 bombs, and four drop tanks were loaded.

Iraq's chemical capability

Chemical weapons were the first WMDs produced by Iraq. Mustard gas had been produced on a large scale by 1983; the nerve agent tabun was produced and weaponized by 1984 and the nerve agent sarin by 1987. Iraq used chemical weapons, including mustard gas and the nerve agents tabun and sarin, extensively against Iranian forces during the Iran–Iraq war between 1982 and 1988. Iraq believed that its use of chemical weapons helped balance Iran's advantage in manpower and persuaded Iran to agree to a ceasefire. At the end of the war, Iraq had the largest and most advanced chemical weapon capability in the Middle East.

At the end of the war, Iraq had the largest and most advanced chemical weapon capability in the Middle East

Iraq's main chemical weapon production facility was at Al Muthanna, also known as Samarra, which began operating in 1983. It produced mustard gas, and the nerve agents tabun, sarin and VX. The plant also produced and filled chemical-warfare munitions. The munitions included missile warheads, aircraft bombs, artillery shells and rockets. The total number of munitions produced ran into tens of thousands.

North Korea

North Korea is an extraordinarily closed and secretive country. It began operating a small nuclear research reactor in 1965 at the Yongbyon nuclear facility, capable of producing plutonium for nuclear weapons. The facilities at Yongbyon include a small reprocessing plant to remove plutonium from spent reactor fuel elements, a plant to make reactor fuel elements, and two partially built nuclear power reactors.

North Korea was suspected of having an active nuclear-weapon program up to 1994. In 1994, it signed an agreement with the USA in Geneva to stop all its nuclear activities. In exchange, the North Koreans were to receive an annual delivery of 500,000 tons of heavy fuel oil and two new nuclear-power reactors, scheduled for completion in 2003 but later put back until 2008. These reactors would be less suitable for producing plutonium for use in nuclear weapons than North Korea's own Yongbyon reactor. The IAEA was to inspect North Korea's nuclear facilities to ensure that the agreement was not being violated.

In November 2002, the USA suspended the oil shipments to North Korea because North Korea would not agree to halt its nuclear weapon ambitions. Soon afterwards, North Korea announced that it had reactivated the nuclear facilities that were mothballed in 1994. In December, Pyongyang reportedly moved fuel rods to the Yongbyon reactor and technicians began work to restart the reactor. At this time, North Korea ordered two IAEA inspectors to leave the country. When operating, the reactor can again produce plutonium for nuclear weapons.

Before 1994, North Korea removed spent fuel elements from the reactor and apparently reprocessed them. Because of these events, North Korea is believed to have nuclear weapons already— American intelligence says two of them. Spent fuel rods that were

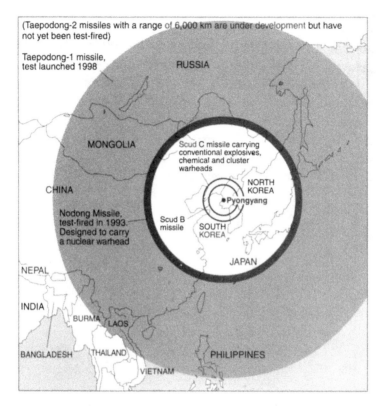

(Taepodong-2 missiles with a range of 6,000 km are under development but have not yet been test-fired)

Taepodong-1 missile, test launched 1998

RUSSIA

MONGOLIA

Scud C missile carrying conventional explosives, chemical and cluster warheads

NORTH KOREA

Pyongyang

CHINA

Nodong Missile, test-fired in 1993. Designed to carry a nuclear warhead

Scud B missile

SOUTH KOREA

JAPAN

NEPAL

INDIA

BURMA LAOS

BANGLADESH THAILAND

VIETNAM

PHILIPPINES

North Korea's ballistic missile capability

put into storage in 1994 could also be used to extract plutonium for perhaps three or four more weapons.

According to US accounts, at a meeting in October 2002 with US Assistant Secretary of State James Kelly in Pyongyang, the North Koreans admitted that they are actively pursuing a nuclear weapons program. Publicly, North Korea has said that it retains "the right" to have nuclear weapons. In the absence of IAEA inspections, the outside world simply does not know how advanced North Korea's nuclear-weapon program is. It is suspected that over the years China and Pakistan have helped North Korea to develop its program.

Pyongyang told James Kelly that it was making separate efforts to produce enriched uranium, presumably as part of its nuclear-weapon program. According to America's CIA, the uranium-enrichment program could be producing two or more nuclear weapons a year by about 2006. The

North Korea has said that it retains "'the right" to have nuclear weapons

pressing question is, of course, will Pyongyang eventually sell nuclear material and technology to other countries, particularly in the Middle East, as it now sells ballistic missiles?

6 What is the international impact of a WMD program?

The further spread of WMDs, particularly in regions regarded as unstable, is widely regarded as a serious threat to regional and global security. Since September 11, 2001, the risk that terrorist groups will acquire WMDs is seen by some as an even greater threat to security than proliferation to states. Of course, the two are frequently linked; some countries could, it is alleged, give terrorist groups WMDs or the materials with which to make them.

Political leaders, Tony Blair and George W. Bush in particular, argue that it is the character of the political leader rather than of the country itself that determines how much of a threat the acquisition of WMDs poses. Some countries in stable regions may acquire WMDs without becoming a serious threat to their neighbors. Diplomatic action may be taken to persuade them to give up their WMDs but no stronger action is likely to be necessary. On the other hand, a megalomaniacal, repressive dictator, with ambitions to increase his power in his region, who therefore poses a threat to his neighbors, is an unacceptable menace if he acquires WMDs. Preemptive military action, American political leaders say, may then be necessary to disarm him. The risk that he might pass on WMDs to a terrorist group is unacceptably high.

There is considerably more concern about the proliferation of nuclear weapons than about biological or chemical weapons. Nuclear weapons are believed to be a much more effective deterrent and to confer more regional status on their owners, and biological and chemical weapons are much less threatening to military forces. In the 1991 Gulf War, for example, Iraq's biological and chemical weapons had relatively little impact on the military tactics of the coalition forces. If Iraq had had nuclear weapons, it would have been a different story.

Because of the different attitude to nuclear weapons, countries that have nuclear weapons are keen to prevent, or least limit, their proliferation to those that do not by attempting to strengthen the existing nonproliferation regime. The suspicion that a new country has ambitions to acquire nuclear weapons focuses attention on the need to strengthen the regime.

Nuclear proliferation

The Nuclear Non-Proliferation Treaty (NPT)

The key international instrument to prevent the proliferation of nuclear weapons is the 1970 Nuclear Non-Proliferation Treaty. Other important nuclear nonproliferation measures include the establishment of zones free of nuclear weapons and the control of the export of nuclear facilities and materials.

As of January 1, 2002, 188 countries had ratified the NPT, making it the most comprehensive multilateral arms control treaty ever. The NPT commits the established nuclear-weapon parties (defined in the treaty as the powers that had manufactured and exploded a nuclear weapon before January 1, 1967), China, France, Russia, the UK and the USA, not to transfer nuclear weapons and not to assist in their manufacture by the non–nuclear weapon states. It also commits the non–nuclear weapon states not to receive nuclear weapons or assistance in the manufacture of them. To verify compliance with the treaty, the non–nuclear weapon parties must sign agreements with the IAEA submitting all their nuclear activities to IAEA safeguards. To encourage the non–nuclear weapon states to join the NPT, the treaty promises cooperation and assistance to these countries in their civil nuclear programs. The treaty obliges China, France, Russia, the UK, and the USA to take significant steps towards nuclear disarmament.

Iraq is a party to the NPT. It was, therefore, regularly inspected by the IAEA. Iraq's nuclear-weapon program before 1991 was an illegal activity, a violation of its treaty obligations. The fact that the IAEA failed to detect Iraq's nuclear-weapon activities has raised questions about the effectiveness of the NPT's verification measures; Iraq's nuclear program was unknown until a defector told the Americans about it. On the other hand, the expulsion of IAEA inspectors by North Korea, a party to the NPT until its withdrawal in January

The NPT has little impact on a "rogue" state intent on developing nuclear weapons

2003, has alerted the world community to North Korea's ambitions to acquire nuclear weapons. The world would probably have not known about North Korea's intentions and capabilities were it not for the country's membership of the NPT.

The cases of Iraq, North Korea, and perhaps Iran, which may have started a program to develop nuclear weapons, show that a country intent on acquiring nuclear weapons can establish such a program while a party to the NPT, taking advantage of its membership of the treaty to obtain assistance in acquiring nuclear expertise and technology. The NPT has little impact on a "rogue" state intent on developing nuclear weapons. The realization of this situation, i.e., the discovery of a clandestine nuclear-weapon program has a significant international impact.

Israel, India and Pakistan are other important countries which are outside the NPT. One reason why they will not ratify the NPT is that they doubt the effectiveness of IAEA safeguards to verify compliance with the treaty. They cannot, therefore, be sure that a

Countries outside the Non-Proliferation Treaty

potential adversary, even though a party to the NPT, will be prevented from developing nuclear weapons. In practice, the NPT depends on the parties to the treaty acting legally and fulfilling their obligations under the treaty. Parties prepared to behave illegally can violate the treaty and establish a clandestine nuclear-weapon program.

Nuclear export controls

The NPT and international safeguards have been shown to be insufficient to prevent countries acquiring nuclear weapons. To bolster these measures, the major exporters of nuclear technology and materials which could be used in a nuclear-weapon program have established guidelines on nuclear exports and some countries have adopted unilaterally national policies to prevent the export of sensitive technologies and materials. These export controls are applied particularly to countries who are not parties to the NPT or who have not accepted IAEA safeguards on all their nuclear facilities.

The major suppliers of nuclear technology and materials have set up the Nuclear Supplier Group, which agreed a list of sensitive materials, equipment, and technology and meets from time to time to review and update the list. These guidelines are not legally enforceable and there are no sanctions for violating the guidelines. They have not prevented some importers using nuclear technology and materials in nuclear-weapon programs. Iraq is a good example.

Weaknesses in the existing international and national measures to prevent nuclear proliferation are one reason why the Bush administration, provoked by the September 11 terrorist attacks, has adopted a policy of taking unilateral and preemptive action against some regimes that develop WMD programs and which may help a terrorist group acquire them. The current American administration, in any case, prefers to act unilaterally rather than multilaterally.

Biological proliferation

Although many countries recognize the dangers and risks of the proliferation of biological weapons, international attempts to control them have not created a regime as effective as those established for nuclear and chemical weapons. In fact, the regime to control biological weapons is a very weak one. This is mainly due to pressures from civil biotechnology and pharmaceutical industries to prevent the implementation of measures to inspect plants that may be involved in the production of biological-warfare agents and munitions. The civil industries fear that the inspections will be a cover for industrial espionage. The highly competitive industries insist on strict commercial confidentiality that is seen to be incompatible with verification activities.

Popular revulsion at the use of chemical weapons, aroused by their use during the First World War, stimulated political leaders to try to ban their use. The Biological and Toxin Weapons Convention (BWC) was opened for signature in 1972 and came into force in 1975. By January 1, 2002, 145 countries had ratified the Convention and eighteen more had signed it. Israel is a country that has neither signed nor ratified the BWC; Egypt and Syria have signed but not ratified it. Signing a treaty implies that the country intends to abide by it, but only ratification makes the treaty binding under the country's laws.

The BWC is the first treaty to ban an entire class of weapons

The BWC bans the development, production and stockpiling of biological and toxin weapons and requires the "destruction of the agents, toxins, weapons, equipment and means of delivery in the possession of the parties." It is the first treaty to ban an entire class of weapons. The treaty, however, contains no provision for checking that the parties are obeying the ban. A significant number—probably about eight—of the parties are suspected of developing biological weapons.

Another weakness of the treaty is that the agents banned are those "that have no justification for prophylactic, protective, or other peaceful purpose." This means that uses of biological agents and toxins that can be described as being for peaceful purposes—in medical research for example—are not banned. More worryingly, research into biological warfare is not banned.

Because of these serious—some say fatal—weaknesses in the BWC, efforts are being made to control the export of agents and equipment that could be used to produce biological weapons. These efforts are being coordinated by a group of thirty or so countries, called the Australia Group. Export controls are, in practice, not very effective. In the words of the Royal Society: "a determined aggressor would, if need be, produce BW using unsophisticated equipment not on the lists." Moreover, limiting exports to developing countries that show no interest in developing biological weapons could well hinder their economic interests.

By far the best way of preventing the spread of biological weapons is to strengthen the BWC, particularly by establishing an effective system of verification to make sure that parties are fulfilling their treaty obligations. In spite of the shock of the September 11, 2001 attacks, the USA is still holding out against the establishment of a verification regime because of lobbying by the powerful American biotechnological industries.

Chemical proliferation

Efforts to ban the use of chemical weapons date back to the Hague Conventions of 1899 ("the Contracting Parties agree to abstain from the use of projectiles the object of which is the diffusion of asphyxiating or deleterious gases") and 1907 (which prohibits the use of poison or poisoned arms). After the extensive use of chemical weapons in the First World War, the Geneva Protocol, which entered into force on February 8, 1928, was negotiated on June 17, 1925; the Protocol prohibited "the use in war of asphyxiating, poisonous or other gases, and of bacteriological methods of warfare." But the Protocol, ratified by 133 countries, bans only the first use of chemical weapons. According to most international lawyers, all countries are bound by the Protocol because a ban on the first use of both chemical and biological weapons has become a well-established customary international law, binding on both parties and non-parties.

Growing concern about the spread of chemical weapons provoked the negotiation of the Chemical Weapons Convention (CWC). The Convention was opened for signature on January 13, 1993 and entered into force on April 29, 1997. As of January 1, 2002, 145 countries had ratified the treaty and twenty-nine more had signed it. Iraq, Egypt, Libya, and Syria have not joined the treaty, but Israel has ratified it.

The CWC bans both the use of chemical weapons and the development, production, acquisition, transfer, and stockpiling of chemical weapons. The parties undertake to destroy their chemical weapons and production facilities. They agree not "to assist, encourage, or induce anyone to engage in any activity prohibited" by the Convention.

The Convention does not ban research "directly related to protection against toxic chemicals and to protection against chemical weapons." Research into and production of chemical-warfare agents for "defense" purposes are identical to those carried out for offensive purposes. Allowing research is, many believe, a loophole in the Convention.

Unlike the BWC, the CWC establishes a mechanism for verifying that the parties are not violating their obligations. The Convention establishes a new international agency, the Organization for the Prohibition of Chemical Weapons (OPCW), based in The Hague. The OPCW sends inspectors into chemical plants and other sites in the territory of CWC parties to verify declarations and to ensure that prohibited activities are not taking place. The function of the OPCW in verifying the CWC is similar to that of the IAEA in verifying the NPT: each agency is responsible for verifying that materials are not diverted to, or produced for, nuclear or chemical weapons.

Each party to the CWC "undertakes not to use riot control

agents as a method of warfare." The legality under the CWC of the use of chemical agents in October 2002 to free the hostages held in a Moscow theater by Chechen rebels was, therefore, highly questionable; about 115 hostages and fifty rebels were killed.

Assessing the impact

The impact of the proliferation of WMDs depends to a large extent on the nature of the regime ruling the country which acquires the weapons. If it is a relatively benign regime, even though it may be a dictatorship, and is in a region that is reasonably stable, so that its neighbors do not feel unduly threatened, the impact may be small. But, as events in Iraq in the spring of 2003 have shown, the acquisition of WMDs by regimes ruled by repressive and unpredictable leaders, in sensitive regions, may give rise to military action to find and destroy the weapons and the production facilities. The action may be taken on behalf of the United Nations, a regional organization, or a single power, almost certainly the USA, acting alone or in a coalition. A major impact of the proliferation of WMDs is, therefore, that national sovereignty is no longer sacrosanct. International relations between states have been fundamentally changed.

> *International relations between states have been fundamentally changed*

Dealing with the proliferation of a WMD to a terrorist group is a much more difficult problem. The group is likely to detonate the weapon without warning as soon as it acquires it, to prevent detection by the authorities. It is, therefore, important to make every effort to prevent the group getting the weapon in the first place, by making sure that the materials needed to make biological, chemical or nuclear weapons are strictly controlled and that effective intelligence measures are in place to warn the authorities if a group is in the process of acquiring a WMD.

If these measures fail and a terrorist group acquires a WMD, the impact is likely to be enormous. Political leaders in a democracy may feel it necessary to take draconian methods against the group and any suspected sympathizers. In a democracy, civil liberties may well be damaged or destroyed. Of course, this disruption of the fabric of society may be just what the terrorists wanted in the first place.

PART III

Weapons of mass destruction and terrorism

7 Terrorism with weapons of mass destruction

Concern about terrorism with weapons of mass destruction is not new. As long ago as 1989, George Shultz, then US Secretary of State, warned: "Terrorists' access to chemical weapons is a growing threat to the international community. There are no insurmountable technical obstacles that would prevent terrorist groups from using chemical weapons." At the time, George Shultz believed that Libya might have been producing chemical weapons and might be willing to supply them to terrorist groups.

Shultz's concern was not generally shared at the time, but the situation has since changed dramatically. A major fear of governments today is terrorist attack with biological, chemical, nuclear, or radiological weapons. Moreover, governments warn that such an attack may be imminent and are making preparations to defend against it. In the words of Tony Blair, addressing more than 100 ambassadors and high commissioners, representing Britain abroad, in London on January 8, 2003: "I warn people: it is only a matter of time before terrorists get hold of a WMD." Terrorism, Blair believes, "is a present and urgent danger."

> "it is only a matter of time before terrorists get hold of a WMD"

Until the end of 1991, discussions about the possible use of WMDs by non-state groups were confined to experts, mostly academics, who study terrorism. Nowadays, all aspects of the issue are frequently analyzed throughout the media. An example is the widespread coverage of the preparations made by the authorities to protect against a terrorist smallpox attack.

Given the extensive media coverage of the risk of terrorist attack, it is hardly surprising that it has become a significant concern for everyone as they go about their daily business. Are such fears about mass killings by terrorists justified? Terrorists in Japan have already used chemical weapons in two lethal attacks. It is hard to believe that they will not use WMDs again.

As wars become increasingly destructive, it is perhaps not surprising that terrorists follow suit. To achieve the dramatic effects they seek, they must move to ever-higher levels of violence. The frequent sights on television of great violence in interstate and civil wars, and of violent crime, show them that only extremely violent actions command TV coverage. And in today's world, TV coverage is an essential ingredient of a successful terrorist action. Publicity is "the oxygen" of terrorism.

in today's world, TV coverage is an essential ingredient of a successful terrorist action. Publicity is "the oxygen" of terrorism

Terrorist violence has escalated steadily as time goes on. The destruction of Pan Am flight 103, *Maid of the Seas*, over Lockerbie, Scotland, on December 21, 1988, killing 270 people (243 passengers, 16 crew members and 11 on the ground) was widely seen to be a new level of terrorist violence. The terrorist attacks in New York and Washington on September 11, 2001, killing almost 3,000 people, ratcheted up terrorist violence to a previously unimagined level.

Until the attack on New York's World Trade Center, most leaders of terrorist groups considered that indiscriminate killing of large numbers of people, women and children included, would not further their aims. The events on September 11, 2001 showed that such self-imposed constraints on mass killing no longer apply.

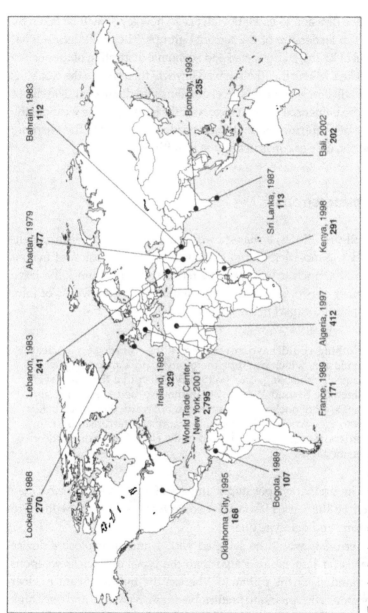

Terrorist death tolls in major incidents

Bahrain, 1983
112

Bombay, 1993
235

Bali, 2002
202

Sri Lanka, 1987
113

Abadan, 1979
477

Kenya, 1998
291

Lebanon, 1983
241

Algeria, 1997
412

Ireland, 1985
329

World Trade Center,
New York, 2001
2,795

France, 1989
171

Lockerbie, 1988
270

Oklahoma City, 1995
168

Bogota, 1989
107

The decision to escalate violence to new levels will be taken by the top leadership of the terrorist groups. They will decide what level of killing and of social and economic disruption of the society attacked will further their aims. They will then choose the weapons that will best achieve this level of killing and disruption. The choice will be influenced by such factors as: the lethality of the weapon; the ease of acquiring the materials needed to fabricate the weapon; and the ease of constructing the munitions.

Nuclear terrorism

Of all the WMDs, a nuclear weapon is, as we have seen, potentially the most lethal and destructive. The explosion of even a primitive nuclear weapon by a terrorist group could kill a large number of people and cause huge damage. In the words of John Despres, an expert in preventing nuclear terrorism:

> Nothing could have anything like the impact of a nuclear explosion, which could be more physically damaging, psychologically shocking, and politically disruptive than any event since the Second World War. Although the casualties from a single act of nuclear terrorism might not match those of a nuclear war, they would still dwarf other forms of terrorism by many orders of magnitude and could easily exceed those of most conventional wars.

The dramatic, apocalyptic impact of a nuclear explosion may well be the very reason why some terrorist groups will try to acquire and detonate one.

Terrorists would be satisfied with a nuclear explosive device that is far less sophisticated than the types of nuclear weapons demanded by the military. Whereas the military want nuclear weapons with precisely predictable explosive yields and very high reliability, most terrorists would not be put off by not being able to predict the power of the explosion.

Now that terrorists have used a chemical weapon in attacks in

Japan, the next level of violence may well be the acquisition and use of a nuclear weapon. A terrorist group may steal a military nuclear weapon, particularly one from the former Soviet arsenal. But it is not only the ex-Soviet nuclear arsenal that we should worry about. A terrorist group may decide to build its own nuclear weapon by acquiring plutonium or highly enriched uranium. It could most easily do so if it could get hold of enough highly enriched uranium.

Luis Alvarez, a leading American nuclear-weapon physicist, has emphasized how easy it would be for terrorists to construct a nuclear explosive with highly enriched uranium: they "would have a good chance of setting off a high-yield explosion simply by dropping one half of the material onto the other half. Most people seem unaware that if highly-enriched uranium is at hand it's a trivial job to set off a nuclear explosion . . . even a high school kid could make a bomb in short order."

Such a primitive gun-type weapon could use a thick-walled cylindrical "barrel," with an inner diameter of about 8 centimeters and a length of about 50 centimeters. A cylindrical mass of highly enriched uranium, enriched to, for example, about 90 percent in uranium-235 and weighing about 15 kilograms, would be placed at the top of the barrel. The larger mass of uranium, weighing about 40 kilograms, would be placed at the bottom of the barrel. This mass would have hollowed out of it a cylinder of the same size as the smaller uranium mass.

A high-explosive charge would be placed at the top of the barrel, behind the smaller mass of uranium. This charge could be fired from a distance by a remote-control device operated by an electronic signal. When the two masses of uranium are brought together, the total mass becomes greater than critical and a nuclear explosion takes place. The gun-type design was used in the nuclear weapon that destroyed Hiroshima.

The total length of the nuclear explosive device is likely to be no more than about 1 meter, and it would be about 25 centimeters in diameter. It should weigh 300 or so kilograms. It could easily be transported by, and detonated in, an ordinary van.

Although highly enriched uranium may be the ideal material for constructing a terrorist nuclear explosive, a terrorist group may find it easier to acquire civil plutonium originally produced in nuclear-power reactors used to generate electricity. After spent reactor fuel is removed from a reactor it is either stored until it can be permanently disposed of in a geological repository or sent to a chemical plant, called a reprocessing plant, where the plutonium in it is separated from unused uranium and the fission products. Because of reprocessing, civil plutonium is becoming more available and it is increasingly possible for a terrorist group to steal or otherwise illegally acquire some. The group could then use the civil plutonium to fabricate a nuclear explosive device.

Of particular concern is the growing trade in civil mixed-oxide (MOX) nuclear fuel. Mixing plutonium oxide with uranium oxide produces MOX. The plutonium oxide is that separated in reprocessing plants from spent nuclear-power reactor fuel elements. MOX is produced in Belgium, France and the UK. It is used to fuel nuclear-power reactors in France, Germany, Sweden and Switzerland. Japan plans to use MOX fuel in its nuclear-power reactors. MOX is, therefore, transported from France and the UK to Germany, Sweden and Switzerland and will be transported from France and the UK to Japan.

If terrorists acquire MOX fuel, they could relatively easily remove the plutonium oxide from it chemically and use it to fabricate a nuclear weapon. The global trade in MOX, therefore, considerably increases the risk of nuclear terrorism. The storage and fabrication of MOX fuel assemblies, their transportation and storage at nuclear-power stations on a scale envisaged by the nuclear industry will be extremely difficult to protect. The risk of diversion or theft of fuel pellets or whole fuel assemblies by personnel within the industry or by armed and organized terrorist groups is a terrifying possibility.

The operators of the nuclear-power reactors that use MOX fuel may want to send their spent MOX fuel elements for reprocessing. They therefore demand that the spent MOX fuel can be dissolved in nitric acid for ease of reprocessing. This requirement makes it much easier for terrorists to separate chemically the plutonium from uranium in MOX.

The chemical separation of plutonium from uranium in MOX fuel pellets is facilitated by the fact that these elements have very different chemistries. The procedures required would be simple and well within the technological capabilities of a moderately sophisticated terrorist organization. The preparation, by the AUM group, of sarin for the attack on the Tokyo underground involved considerably more sophisticated chemistry and greater acute danger to the operators than that required for the separation of plutonium from MOX; the chemistry is less sophisticated than that required for the illicit preparation of designer drugs.

the chemistry is less sophisticated than that required for the illicit preparation of designer drugs

None of the concepts involved in understanding how to separate the plutonium are difficult; a second-year undergraduate might be able to devise a suitable procedure by reading standard reference works, consulting the open literature in scientific journals, and by searching the World Wide Web. The progress of the separation can be estimated easily at different stages by measuring the concentrations of uranium and plutonium, by, for example, ultraviolet spectrophotometry, using cheap and readily available equipment.

The simplest method of separating the plutonium and uranium in MOX involves an ion-exchange column filled with a resin, a standard piece of chemical apparatus readily acquired. When properly used, an ion-exchange method using a suitable resin gives excellent and rapid separation with better than 92 percent efficiency. Suitable resins are extensively used in ion-exchange columns by industry for water softening, waste treatment, and resource recovery.

The resins, of which there are several different types, can be easily purchased "off-the-shelf" in large quantities with short delivery dates.

If the terrorists are not satisfied with the percentage of the plutonium separated from the MOX by the first run through their ion-exchange column, they could simply repeat the cycle. This should yield plutonium with a purity of at least 99 per cent.

A primitive nuclear explosive constructed from plutonium oxide separated from MOX may have an explosive yield smaller and less predictable than a device constructed from plutonium metal. The terrorists may, therefore, decide to prepare plutonium metal from the plutonium oxide they have separated from MOX nuclear fuel. This can be done using standard chemical techniques.

Terrorists may use plutonium oxide obtained from MOX directly or convert it chemically into plutonium metal, a straightforward process. In either case, a sphere of the material would be used and a supercritical mass produced using a technique called implosion. (The gun-type design cannot be used with plutonium.)

The sphere of plutonium is surrounded by conventional high explosives. When exploded, the high explosive uniformly compresses the sphere, reducing its volume and, therefore, increasing its density. The critical mass is inversely proportional to the square of the density. The original less-than-critical mass of plutonium will, after compression, become supercritical, causing a nuclear explosion. The critical mass of plutonium oxide is greater than that of the metal—about 35 kilograms compared with about 13 kilograms. A terrorist group prepared to convert the plutonium oxide to the metal would, therefore, need to acquire significantly less plutonium oxide.

A sphere of civil plutonium oxide having a critical mass would be about 18 centimeters in diameter; a sphere of plutonium metal having a critical mass would be about 6 centimeters in diameter. If the plutonium sphere is surrounded by a shell of material, such as beryllium or uranium, neutrons that escape from the sphere without producing a fission event are reflected back into the sphere. A reflector, therefore, reduces the critical mass. A thick reflector will reduce the critical mass by a factor of two or more.

The high explosive could be TNT or RDX. But it is more likely that a terrorist group would use a plastic explosive, such as

Semtex, since it is easier to handle and can be molded into a spherical shape around the plutonium sphere to ensure more even compression of the plutonium. About 400 kilograms of plastic explosive, molded around the reflector placed around the sphere of plutonium, should be sufficient to compress the plutonium enough to produce a nuclear explosion.

If a 5-centimeter thick shell of beryllium was used as the reflector and surrounded by the 400-kilogram shell of plastic explosive, the assembled device would have a radius of about 40 centimeters constructed from 18 kilograms of plutonium oxide (a sphere with a radius 7.3 centimeters), the beryllium reflector, and the plastic explosive. If 7 kilograms of plutonium metal was used instead, and the plutonium sphere (radius 4.8 centimeters) was surrounded by a 5-centimeter shell of beryllium and 400 kilograms of plastic explosive, the radius of the total device would be just less than 40 centimeters.

The size of the explosion from such a crude device is impossible to predict. But even if it were only equivalent to the explosion of a few tens of tons of TNT it would completely devastate the center of a large city. Such a device would, however, have a strong chance of exploding with an explosive power of at least 100 tons of TNT. Even 1,000 tons or more equivalent is possible, but unlikely because the compression achieved in a primitive design will probably not be symmetrical enough to produce such a large nuclear explosion.

Even if a primitive nuclear weapon using plutonium, when detonated, did not produce a significant nuclear explosion, the explosion of the chemical high explosives would disperse the plutonium widely. If an incendiary material, such as an aluminium-iron oxide (thermite), were mixed with the high explosives, the explosion would be accompanied by a fierce fire. A high proportion of the plutonium is likely to remain unfissioned and would be dispersed by the explosion or volatilized by the fierce heat. Much of the plutonium is likely to be dispersed in this way as small particles of plutonium oxide taken up into the atmosphere in the fireball and scattered far and wide downwind.

A large fraction of the particles are likely to be smaller than three microns in diameter, and could therefore be breathed into, and retained by, the lung. Here they would be very likely to cause lung cancer by irradiating the surrounding tissue with alpha particles. This is why inhaled plutonium is so highly toxic.

Once dispersed into the environment, plutonium oxide is insoluble in rainwater and would remain in surface dusts and soils for a protracted period. The half-life of the plutonium isotope Pu-239, the predominant form in civilian plutonium, is 24,400' years so that it will, in effect as far as humans are concerned, stay radioactive forever.

These factors would combine to render a large part of the city uninhabitable until decontaminated by washing down buildings, cleaning road surfaces, removing topsoil, and so on. Decontamination could take many months or even years. The threat of dispersion of many kilograms of plutonium makes a crude nuclear explosive device a particularly attractive weapon for a terrorist group, the threat being enhanced by the general population's justifiable fear of radioactivity.

The sheer amount of plutonium in the world is itself an incitement to nuclear terrorism

The sheer amount of plutonium in the world is itself an incitement to nuclear terrorism. Plutonium was first discovered in 1940 and, as we have seen, first produced in significant amounts as part of the Manhattan project, set up by the Americans in the Second World War to manufacture nuclear weapons. Plutonium, which has a silvery look, is radioactive because of the energy released during radioactive decay and a fairly large piece is warm to the touch; a large piece will produce enough heat to boil water. It is a very heavy material, with a density nearly twice that of lead. Since 1945, the world has produced a huge amount of plutonium—a total of about 1,500 tons.

About 250 tons of this plutonium was produced for military use in nuclear weapons. The other 1,250 tons are civilian plutonium produced as an inevitable by-product by civilian nuclear-power reactors while they are generating electricity.

About 300 tons of civil plutonium have been separated from spent nuclear-power reactor fuel elements in reprocessing plants; on current reprocessing plans, about 550 tons of civil plutonium will be separated by the year 2010. About 20,000 nuclear weapons could be fabricated from the 300 or so tons of separated civil plutonium in the world today.

About 80 tons of this civil plutonium are currently stored in France, about 60 tons in the UK, about 50 tons in Japan, and about 40 tons in each of Germany and Russia. Smaller amounts (less than 8 tons) are in each of Belgium, India, Italy, the Netherlands, Spain, Switzerland, and the USA.

The situation with highly enriched uranium is different from that with plutonium. The bulk of the world's stock of highly enriched uranium, about 99 percent, is military. Moreover, surplus highly enriched uranium can be disposed of by simply mixing it with natural to make it unusable as a nuclear explosive. Military uranium is probably kept more securely than civil plutonium and is, therefore, less easily acquired by terrorists.

Biological and chemical terrorism

Biological and chemical weapons are in a different category from nuclear weapons. In most circumstances, nuclear terrorism is likely to be much more lethal and destructive than either biological or chemical terrorism. But the materials required to fabricate nuclear explosives are harder to come by than the materials needed to make biological or chemical weapons.

Of all types of WMDs, the materials and equipment needed to produce chemical weapons are easiest to acquire. The chemicals and chemical equipment needed to prepare tabun, the easiest nerve agent to produce, are relatively straightforward to obtain. Methods of manufacturing nerve agents are described in the open literature and the original papers describing them are not difficult to find. References to these papers can be found in a good public library, and a competent chemist would have little difficulty in preparing nerve agents.

Biological agents could be stolen or taken out by a sympathizer from, for example, a medical research institution. Alternatively, agents such as anthrax bacteria or botulinum toxin can be obtained from natural sources. On balance, though, a chemical weapon based on tabun would probably be easier to fabricate than a biological one.

It is, however, likely to be less lethal. Even if a terrorist group disperses a chemical weapon—a nerve agent for example—using effective technology, the lethal effects will be local, rather than widespread. People have actually to inhale the nerve agent or absorb it through the skin. Contagious biological agents, like smallpox or plague, on the other hand, can be passed from one individual to another, a process that can spread the disease over a relatively large area. Terrorists are, therefore, likely to kill more people with biological weapons than chemical ones.

Nevertheless, a terrorist group would probably choose a chemical weapon based on a nerve agent as its first WMD. The technology for dispersing a chemical-warfare agent such as a nerve agent is not difficult to acquire or master. The obvious technique for dispersal is the production of an aerosol to release the agent as a cloud of droplets.

Although chemical weapons are, on balance, easier to make than biological ones, populations fear a biological attack more than a chemical attack. Official concern about bioterrorism dates back to the discovery of the AUM group's program to develop biological weapons. Public fear of bioterrorism grew considerably after the anthrax letter attacks in the United States that followed September 11, 2001.

In an article in the magazine *Scientific American*, Leonard A. Cole explains the fear of bioterrorism:

> If a chemical attack is frightening, a biological weapon poses a
> worse nightmare. Chemical agents are inanimate, but bacteria,

viruses and other live agents may be contagious and reproductive. If they become established in the environment, they may multiply. Unlike any other weapon, they can become more dangerous over time.

Terrorists may see this psychological effect as a significant advantage for bioterrorism.

Terrorists have been found in possession of biological agents on a number of occasions. In 1972, for example, members of the right-wing group, the Order of the Rising Sun, were arrested in Chicago with about 35 kilograms of typhoid bacteria cultures. The terrorists intended to poison water supplies in Chicago, St. Louis, and other cities.

In 1984, members of the Rajneesh cult contaminated salad bars in the Oregon restaurant, The Dalles, with bacteria that cause typhoid fever. Seven hundred and fifty people became ill although none died. In the 1980s, a large amount of botulinum toxin was found in a Parisian house used by Red Army Faction terrorists.

members of the Rajneesh cult contaminated salad bars with bacteria that cause typhoid fever

And the AUM group cultured anthrax bacteria in drums of liquid in the basement of its eight-story building in Kameido, a Tokyo suburb. The liquid was pumped to the roof and sprayed into the air for twenty-four hours.

There were no reported symptoms of anthrax and it was first assumed that the attack had failed. But it was later discovered that the anthrax used by the AUM group was of the Sterne strain that does not cause anthrax in humans. The AUM group may have used the harmless anthrax bacteria to practice their techniques before moving on to virulent ones. The attention of the police may have discouraged them from this further action.

It will surprise many to learn that biological agents would not have to be acquired from civilian or military research laboratories, from a sympathizer working in the laboratory or by theft: they can be bought from legitimate suppliers. This is shown by the case of Larry Harris, a member of the white supremacist group Aryan Nations. Harris telephoned the American Type Culture Collection in Maryland and ordered three vials of freeze-dried bacteria that cause bubonic plague. While they were in transit with Federal Express, he called the supplier to ask where they were. The supplier became suspicious about his impatience and contacted the authorities. Harris was charged with mail fraud. Had he quietly waited for the delivery of the bacteria he would not have been caught.

The operations needed by a terrorist group to fabricate a WMD would require a degree of sophistication, but terrorist organizations are certainly capable of sophisticated planning and the application of scientific principles. Ruthless terrorists are not likely to be concerned about their own safety or about contaminating the environment with chemical, biological, or radioactive material; although they will avoid environmental pollution, through accidents or releases, that might reveal their clandestine activity.

A well-funded terrorist group would not be unduly constrained by cost, would have or could employ the services of specialists with relevant experience, and would have access to standard laboratory equipment.

8 Which groups are capable of making and using a WMD?

The nuclear terrorist

A terrorist group intent on making a primitive nuclear explosive needs plutonium or highly enriched uranium. Both of these materials are toxic; they are poisonous and emit radiations that are damaging to human health. They have a particularly high inhalation toxicity and so must be handled with care.

A group of five senior American nuclear-weapon designers addressed the question: could a terrorist group make a nuclear explosive? They concluded that it could but pointed out the potential hazards, which included: "those arising in the handling of a high explosive; the possibility of inadvertently inducing a critical configuration of the fissile material at some stage in the procedure; and the chemical toxicity or radiological hazards inherent in the materials used."

A "critical configuration of the fissile material" would occur if enough of the material accumulates under conditions to produce a significant number of fissions. Neutrons would be produced by such an event. Any person in the vicinity would be irradiated with neutrons, a potentially very hazardous experience.

An American nuclear physicist, Amory Lovins, commenting on the views of the nuclear-weapon designers, argues that these hazards should not be exaggerated. He shows that the radiation dose to which the terrorists may be exposed when handling the uranium and plutonium would be unlikely to deter them. And he concludes that, given sensible precautions against achieving criticality accidentally (for example, making sure that not too many of the materials are accumulated in one container) a terrorist group could avoid serious hazards. A neutron counter could be used to detect any neutrons emitted during the assembly of the plutonium. An increase in the number of neutrons escaping from the plutonium would indicate an approach to criticality. A terrorist group constructing a nuclear explosive would thus not face serious radiological hazards. In any case, such a group would probably be prepared to take some risks to achieve their purposes.

A terrorist group may use plutonium metal, plutonium oxide, highly enriched uranium oxide, or highly enriched uranium metal as the fissile material in a primitive nuclear explosive. The terrorists could considerably reduce the mass of fissile material required by surrounding it with a thick neutron reflector. If they acquired either plutonium oxide or uranium oxide, they would probably decide to convert it to the metal because they would need much less of the oxide. The preparation of the metal from the oxide is a straightforward chemical procedure.

One reason to convert the oxide to the metal is to avoid the rather awkward task of compacting it to obtain the highest possible density. The greater the density, the lower the critical mass. The compaction will require the use of a large and special press. The group may prefer not to risk drawing attention to it by acquiring a suitable one. The other equipment needed would be standard and the acquisition of it is not likely to attract unwelcome interest.

The team of terrorists setting out to fabricate a nuclear explosive will include at least one competent graduate nuclear physicist and at least one graduate chemist with experience in working with uranium or plutonium compounds. The team will probably also include a person who knows how to handle conventional high explosives safely and an electronics expert.

If the nuclear explosive being built by the terrorist group uses implosion to compress the fissile material to obtain a critical mass, the member competent in handling explosives will arrange the high explosives around the fissile material and the electronics person will arrange a circuit to fire the detonators to set off the high explosive.

Work with plutonium or uranium must be done in a fume cupboard to reduce the risk that the substance will be inhaled or ingested. A simple structure would be sufficient. This could be a sealed plastic box within the room, with a chimney to the outside, provided with rubber gloves through which materials and equipment could be handled.

The bioterrorist

To fabricate a biological weapon a group must first obtain an adequate quantity of an appropriate strain of the chosen biological-warfare agent; it must then acquire the technology to disperse the agent effectively. The group must have among its members people with the knowledge of how to handle the strain correctly and safely, how to grow

Bacteria can be grown in artificial media using methods similar to those used in the brewery industry

(culture) the strain in an appropriate way, and how to store it properly. Bacteria can be grown in artificial media using methods similar to those used in the brewery industry. Viruses and rickettsiae must be grown on living tissue.

If the group decides to obtain a biological agent from nature—soil, contaminated food, or the corpses of animals that have died from disease—it will isolate the agent and culture it. To grow a biological agent the group will again need to set up a laboratory containing fermenters. The facilities and equipment needed to produce adequate amounts of a biological agent are generally relatively cheap and unsophisticated. A competent biologist with knowledge of growth media will be employed in the laboratory.

A reasonably sophisticated terrorist group with access to financial and technical resources and able to employ people adequately trained in biology and in handling biological material will be able to acquire a biological-warfare agent and establish a program to produce an effective biological weapon using it.

A way of lengthening the life of bacteria such as anthrax is to freeze-dry them, a technique that freezes the bacteria so rapidly that the formation of ice crystals in the bacteria is inhibited. The bacteria must then be reconstituted before dissemination.

Freeze-drying is not an easy technique to master; more difficult than any other technique in a biological-weapon program. Terrorists are, however, unlikely to freeze-dry bacteria because they will probably want to use the biological agent very soon after they acquire it. They will not want to keep the agent around long in case the authorities find it. The aim of the terrorists will be to get the weapon and use it as soon as possible, so they are likely to have the dispenser ready before acquiring the agent. They will then load the agent into the dispenser and use the weapon quickly.

The technology required to disperse a biological agent effectively is not difficult to master, well within the capability of a group with significant resources of money and qualified people. In most cases, the group will probably disperse the agent as an aerosol.

The chemical terrorist

A modern terrorist group planning to use a chemical weapon will almost certainly choose to fabricate a nerve agent. To do so, it will need to set up a chemical laboratory with adequate fume cupboard facilities. Because of the toxicity of nerve agents, fume cupboards will be provided with negative pressure so that if there is a leak no gas can escape from the cupboard into the laboratory.

Before removal from a fume cupboard, the nerve agent, a liquid, will be contained in an airtight container until it is loaded into a disperser, probably an aerosol generator. Nerve agents are generally volatile liquids so that very primitive methods of dispersal could be used. The AUM group contained their sarin nerve agent in a plastic container that they punctured with an umbrella with a sharpened tip. The group would, however, have killed more people if they had used an aerosol.

The agent could, for example, be injected into the air-conditioning system of an office block or dropped into a tunnel of an underground train system. In an open-plan building, a vial of a nerve agent could be broken at a position that would ensure contamination of the air in the building.

None of the chemicals involved in the chemical preparation of, for example, the nerve agent tabun (dimethylamine, phosphoryl chloride, sodium cyanide, and ethanol) are themselves particularly toxic. Great care must be taken when handling the nerve agent itself, however, though no exceptional precautions are necessary until it is produced.

The prime suspects

Until the 1970s, terrorist groups were generally fighting for the independence of their countries or regions from colonial rule. Since then, however, most terrorism has arisen from religious, political, ethnic, cultural, nationalistic, and ideological conflicts. The motivation of many modern groups consists of more than one of these elements.

Bearing in mind that unambiguous dividing lines cannot usually be drawn, there are five types of terrorism, each with distinct characteristics: terrorism by an individual; political terrorism, usually with nationalist aims; terrorism by extreme political groups, right- and left-wing; terrorism carried out by single issue groups, such as antiabortionists and radical ecologists; and terrorism by religious fundamentalists.

An act of violence by one person can be called an act of terrorism if the individual believes he/she is acting for a cause or if his/her aim is to "change society." A salient example of a terrorist act by an individual is the assassination in November 1995 of Israel's Prime Minister Yitzhak Rabin. The assassin was an Israeli law student, Yigal Amir, a member of the extreme right-wing group Eyal, who said that he killed Rabin because the Prime Minister's plan to give up Israeli territory violated religious tenets.

Most terrorist groups with political aims are nationalist; the usual goal is to create an independent homeland. Typical examples are the Basque group ETA, the Tamil Tigers, groups within the Palestinian Liberation Organization (PLO) and groups in Northern Ireland.

Right-wing terrorists generally believe that conflict and violence are essential elements in society. Extreme right-wing groups in America, Eastern Europe, Israel, Japan, and South Africa are closely aligned with religious fundamentalist movements. In Western Europe, however, right-wing extremism is generally not associated with a religion.

> *Right-wing terrorists generally believe that conflict and violence are essential elements in society*

Professor Paul Wilkinson, a leading terrorism expert at St. Andrews University explains that right-wing extremists

> view their enemies not simply as misguided opponents but rather as sub-human, people who should be accorded a subordinate and degrading status, not only legally but in every aspect of life as well. They blame their enemies for all the ills and injustices in society and are willing not only to demonize them and make them into scapegoats and pariahs, but also to countenance the expelling or even the killing of them.

To judge by their actions, religious terrorists are generally the most fanatical of all terrorists. They believe that their terrorist violence is a divine duty, a response to a God-given religious command; religious terrorists are willing, and sometimes even anxious, to die for their cause. In the words of Bruce Hoffman, religious terrorism

> assumes a transcendental dimension, and its perpetrators are
> thereby unconstrained by the political, moral, or practical constraints that seem to affect other terrorists. Whereas secular
> terrorists generally consider indiscriminate violence immoral
> and counterproductive, religious terrorists regard such violence
> not only as morally justified, but as a necessary expedient for the
> attainment of their goals. Thus, religion serves as a legitimizing
> force—conveyed by sacred text or imparted via clerical authorities claiming to speak for the divine.

Hoffman explains that "secular" terrorists regard themselves as
answerable to their supporters, the people they say that they are
fighting for, or the oppressed group they claim to represent.
"Religious terrorists," on the other hand, "execute their terrorist
acts for no audience but themselves. Thus the restraints on violence that are imposed on secular terrorists by the desire to appeal
to a tacitly supportive or committed constituency are not relevant
to the religious terrorist."

Religious terrorists may be prepared to commit almost limitless
violence against almost any target. Any persons who are not members of the terrorist's religion or religious sect are valid targets.
Fundamentalist religious terrorist groups are the most likely to
take the decision to acquire and use a WMD. Who are the religious
fundamentalists?

Islamic Fundamentalists

There are currently two main groups of extreme religious terrorists, the Islamic Fundamentalists and the White Supremacists. The
Islamic groups generally conduct terrorism as a form of Holy War,
to be continued until total victory is won. Some Islamic groups are
"more fundamentalist" than others but they are all unwilling to
compromise. Mullah Hussein Mussawi, the leader of Hizbollah
until he was assassinated in Lebanon by the Israelis in 1994,

explained that Shia Islamic Fundamentalists, for example, are "not fighting so that the enemy recognizes us and offers us something. We are fighting to wipe out the enemy."

Shiites generally believe that secular governments have no legitimate authority. They believe they have an absolute duty to work for the universal implementation of Islamic law as defined in the Koran. The use of very violent acts is an acceptable and, indeed, essential means of fulfilling this duty.

> *"We are fighting to wipe out the enemy"*

Hizbollah (the Party of God), the best-known radical Shia group (also known as Islamic Jihad, the Revolutionary Justice Organization, Organization of the Oppressed on Earth, and Islamic Jihad for the Liberation of Palestine), operates from Lebanon. With several thousand supporters, it has established cells in a number of countries, in Western Europe, Africa, and elsewhere.

Hizbollah was responsible for the bombing of the US Marine Barracks in Beirut on October 23, 1983, by a suicide bomber driving a dump truck carrying about 5 tons of explosives. On the same day, another flatbed truck, carrying nearly 2 tons of explosives, destroyed the French military compound in Beirut. The two bombings killed 298 soldiers.

The nature of fundamental Islamic terrorism has recently changed. Ely Karmon, an expert in the subject, explains that radical Islamist terrorism has

> shifted from the Shia brand, developed under the influence of the Iranian Khomeinist revolution, to the Sunni model (that emphasizes consensus and community). The significance of this is in the fact that the Sunnis are in an overwhelming majority over Shias in the Muslim world. Thus, the threat from this kind of terrorism has grown, and we now see large countries like India, China and Russia confronting this kind of terrorism. For this reason also, the volume of terrorism has sharply increased in many Arab and Muslim countries—Algeria, Egypt, Pakistan, Saudi Arabia, Caucasus, Central Asia, and lately Indonesia—in the Israeli-Palestinian conflict, and in countries with Muslim

minorities, such as the Philippines. The third consequence of the Shia-to-Sunni shift is the appearance of international networks of Islamist terrorists, the most famous being Osama bin Laden's Al Qaeda.

White Supremacists

The other main radical religious terrorist groups are made up of Christian White Supremacists, part of the radical right wing in the United States. Like the Islamic Fundamentalists, the Christian White Supremacists justify and legitimize terrorist violence on the basis of their religious beliefs. White Supremacist groups preach a virulent anti-Semitism vilifying Jews and non-whites as the children of Satan.

A number of these groups are involved in the white supremacist movement known as Christian Identity, including a number of influential Identity churches across the USA, the modern version of which emerged in the 1940s. The most important of them are Aryan Nations and the Order, also known as the Silent Brotherhood, a splinter group of the Aryan Nations. Many of the members of these groups have a messianic belief in the Second Coming of Christ, although, of course, this time a white Aryan Christ.

Members of Christian Identity come mainly from conservative Protestant churches. Christian Identity shares with Protestant fundamentalism an apocalyptic belief but with a crucial difference. Jeffrey Kaplan, a leading expert in right-wing violence in the United States, describes this difference:

> where fundamentalists can await the eschatological "End of Days" secure in the knowledge that in the dreaded seven-year period of the Tribulation when war and famine and disease engulf the earth they will be raptured into the air to await the inevitable conclusion of history at Jesus' side, the Identity believer has no such hope of supernatural rescue. Rather, the Christian Identity believer is secure only in his ability to persevere—to survive by the grace of God, by virtue of his own wits and through recourse to his own food stores and weapons.

The White Supremacists emphasize the importance of survivalism, of the acquisition and maintenance of arsenals of weapons and of training in the use of the weapons. To prepare for Armageddon some of them are accumulating stocks of food and weapons and training themselves to survive the holocaust.

Christian White Supremacists are generally opposed to any form of government above the level of local government. They believe that Jews control the American government, financial centres and the media. They call the American government the "Zionist Occupation Government" or ZOG; their main aim is to overthrow the ZOG. Some White Supremacists believe that any level of violence is justified to destroy the groups they hate and in their war against the ZOG.

To prepare for Armageddon some of them are accumulating stocks of food and weapons and training themselves to survive the holocaust

Many members of America's extreme right-wing groups have been strongly influenced by Dr. William Pierce, an ex-physics professor at Oregon State University and a guru of the far right. His book *The Turner Diaries*, written under the pseudonym of Andrew MacDonald and published in 1985, is an apocalyptic novel describing the activities of the hero, Earl Turner, the leader of a terrorist group, called the Order, which wages a terrorist race war against a Jewish-controlled American government. The book contains a detailed account of the bombing of a Federal building with a fertilizer bomb. The idea to bomb the Murrah building in Oklahoma with a fertilizer bomb may have come from this account.

At the end of the book, terrorists belonging to the Order capture America's nuclear weapons and use them to destroy a number of American cities and then to attack Israel and the former Soviet Union. *The Turner Diaries* is avidly read by white supremacists in the United States.

The increased threat

The frequency of international terrorist attacks and their increasing lethality are causes of great concern. The number of international terrorist attacks over the past twenty years has varied between 274 and 666 a year, with a yearly average of 459. The lowest number, 274, occurred in 1998 but these attacks killed and wounded a record number of people, with 741 people killed and 5,952 injured. The record total arose from the bombings of the American Embassies in Nairobi, Kenya, and Dar es Salaam, Tanzania. The first killed 291 persons and wounded about 5,000; the second killed 10 and wounded 77. Osama Bin Laden, his military commander Muhammad Atef, and members of the Al Qaeda terrorist group—a total of twenty-two persons—were charged with the bombings.

The lethality of international terrorist violence has increased dramatically over the past thirty years. Between 1995 and 2000, for example, a total of about 20,000 people were killed and injured in international terrorist attacks. There is no reason to believe that terrorist violence will decrease in the foreseeable future.

Al Qaeda and Osama Bin Laden

Since the September 11, 2001 attacks in New York and Washington, Al Qaeda (the Base) has become the most infamous terrorist organization. Al Qaeda, originally called the "International Islamic Front for Jihad against America and Israel," is not an autonomous terrorist group; it is rather an idea or ideology.

With an international network of cells, each containing a number of Islamic fundamentalists with vehement anti-American and anti-Israel ideologies, in ninety countries according to the US State Department and fifty according to the Centre for the Study of Terrorism at St. Andrews University—it is the most dangerous terrorist organization that has existed so far. Its cells are all over the Arab world, in Europe, Asia, the United States, and Canada.

According to a US Congressional report prepared by the

Congressional Research Service, released on September 13, 2001, the Al Qaeda network has access to anti-aircraft missiles and chemical weapons. A number of Islamic terrorist groups, such as the Egyptian Al-Gama'a al-Islamiyya and the Egyptian Islamic Jihad, are thought to be associated with Al Qaeda.

The founder of Al Qaeda was Osama Bin Laden who has become the world's most notorious terrorist. A forty-four-year-old multimillionaire, Osama Bin Laden is the youngest son of a wealthy Saudi businessman. His fortune is said to exceed 300 million US dollars. He became seriously involved in Islamic Fundamentalism in 1979, the year in which troops of the former Soviet Union invaded Afghanistan. He moved his business interests to Afghanistan and organized resistance to the "infidel invader." He developed a worldwide organization to recruit Muslim terrorists to fight against the former Soviets in Afghanistan, running training camps for young Muslims during the Afghan War of the 1980s. His activities against the Soviet occupation of Afghanistan were much encouraged by the Americans.

> *Bin Laden's fortune is said to exceed 300 million US dollars*

Bin Laden and Al Qaeda aim to provoke a war between Islam and the West and to overthrow Muslim governments, such as those of Egypt and Saudi Arabia. Bin Laden justified the formation of the anti-American and anti-Israeli front by arguing that Muslims everywhere in the world were suffering at the hands of the USA and Israel. He said the Muslims must wage holy war against their real enemies not only to rid themselves of unpopular regimes backed by the Americans and Israelis but also to protect their faith. The Bin Laden network is suspected of supporting terrorists in Afghanistan, Bosnia, Chechnya, Tajikistan, Somalia, and Yemen.

He returned to his home in Saudi Arabia in 1989, but the government there expelled him the following year. Bin Laden then went to Sudan from which he carried on his support for terrorist operations. Among his numerous Sudanese commercial interests are: a factory to process goat skins, a construction company, a bank, a sunflower plantation, and an import-export operation.

At the urging of the United States, and following the attempted assassination of President Mubarak of Egypt, in which Bin Laden was involved and in which the Sudanese government was complicit, the government of Sudan expelled Bin Laden in 1996 whereupon he relocated to Afghanistan. However, he has maintained considerable business interests and facilities in Sudan.

On August 20, 1998, the US military struck facilities in Afghanistan and Sudan thought to belong to Osama Bin Laden's network in retaliation for the carefully coordinated attacks, on August 7, 1998, on US embassies in Nairobi, Kenya and Dar es Salaam, Tanzania. Bin Laden and Al Qaeda have also been associated with the killings of Western tourists by militant Islamic groups in Egypt, the 1993 World Trade Center bombing, the 1995 explosion of a car bomb in Riyadh, Saudi Arabia; the 1995 truck bomb in Dhahran, Saudi Arabia; and the motorboat attack on the USS *Cole* off Yemen in October 2000 that killed seventeen American sailors. Allegedly Bin Laden was associated with the attacks on September 11, 2001 on the World Trade Center, New York and the Pentagon, Washington.

Bin Laden has issued three "fatwahs" or religious rulings calling upon Muslims everywhere to take up arms against the United States. One, issued in February 1998, called for the liberation of Muslim holy places in Saudi Arabia and Israel, as well as the death of Americans and their allies, a call for a holy war against Americans. He seems to be motivated by his opposition to the presence of American troops near the holy sites of Mecca, Saudi Arabia, in violation, according to him, of the principle that "the feet of infidels must not sully the Kaaba."

Al Qaeda maintains connections between Muslim extremists in its network, using fax machines, satellite telephones, email and the Internet. Like Bin Laden, Al Qaeda's aims are the overthrow of what it regards as the corrupt and heretical governments of Muslim states, and their replacement with governments that will rule strictly according to Islamic law (Sharia).

In 2001, American troops in Afghanistan, searching caves that had housed Al Qaeda members, found documents describing WMDs, including nuclear explosives. The risk that elements of the organization will in the future acquire WMDs must be taken seriously. The group arrested in London while experimenting with ricin toxin probably contained members of Al Qaeda.

There is a significant risk that Al Qaeda will acquire WMDs; the AUM group has already done so. Learning lessons from the nature and activities of the AUM group will reduce the future risk of the proliferation of WMDs to non-state groups.

The AUM group

The AUM (Supreme Truth) group was created in 1987, at Kamiku Isshiki, a village in the foothills of Mount Fuji, and led by the forty-eight-year-old half-blind Shoko Asahara (whose original name is Chizuo Matsumato), a former yoga instructor and the son of a tatami mat maker. An example of a sophisticated terrorist group with access to large financial, scientific, and technical resources, the AUM group was the first to use an effective chemical weapon. The nature of the group is, therefore, worth discussing in some detail.

The AUM, a doomsday cult, preaches a philosophy combining a blend of Eastern religions with Christianity. Many brilliant young scientists from top universities, particularly Japanese ones, were attracted to the AUM, which also recruited members in Russia, Germany and the United States. The AUM broadcast regularly on Moscow radio and recruited tens of thousands of Russian members. In fact, it appears that the AUM had about 30,000 Russian members compared with about 10,000 in Japan.

In the ten years between its creation and the arrest of Asahara and other leaders of the group in 1996 for the Tokyo nerve gas attack, the AUM established good relations with a number of prominent Japanese politicians, especially with those on the extreme right wing. The group accumulated massive cash reserves, reportedly of some two billion dollars. A great deal of money came from donations made by members of the sect.

The AUM ran many businesses through which it procured weapons and chemicals. The global network set up by the sect was able to procure high-technology equipment and materials from many sources, including some in Silicon Valley, California. After the Tokyo attack the police found tons of chemicals used to synthesize sarin, large quantities of other chemicals, precision machine tools of the types used to manufacture weapons, nitroglycerine, a Russian helicopter, and details about biological-warfare agents and weapons.

Some AUM members were, it is reported, trained by the Russian army in the use of a variety of weapons, including several types of missiles and armoured vehicles. The AUM procured a variety of weapon systems from the Russian army. AUM leaders, including Asahara, made many trips to Russia and established close links with a university in Moscow.

Asahara preached that an apocalyptic war would destroy civilization, after which the AUM would establish its kingdom. The AUM armed itself to prepare for Armageddon, setting its scientists the task of developing powerful weapons, including WMDs.

As part of its program to develop nerve gases, AUM scientists set up a test site in Australia. They arrived at Perth airport with a large amount of chemicals and equipment for a chemical laboratory (it reportedly cost them 330,000 Australian dollars in excess baggage). Some were fined for transporting illegal and dangerous chemicals on a passenger aircraft. The site chosen for the testing ground was located in the outback region of the Leonora-Laverton district of Western Australia. As well as testing nerve agents on sheep (a number of sheep bodies were found by the police), the group was interested in the uranium deposits in sediments in salt lakes. They flew from sheep station to sheep station in a light aircraft, using radiation detectors to prospect for uranium at salt lakes known to have uranium in their sediments. They obtained information about the location of suitable lakes from the Australian Mines Department.

On June 1, 1993 the AUM purchased the half-million-acre Banjawarn sheep station, about 600 kilometers northeast of Perth.

Members, including Asahara, moved into Banjawarn in force during September 1993. Most stayed at the station for a few weeks but some remained until May 1994. Evidence that Banjawarn was used to test sarin nerve gas on a flock of twenty-nine merino sheep was provided by analyses of sheep carcasses.

AUM members were seen by local people and passers-by driving around Banjawarn sheep station, and on roads in the area, dressed in white protective suits, complete with helmets. It is not clear why the AUM scientists set up a base in Australia to test their nerve gas when they could have easily found a suitable site in Japan. They probably did so because they wanted to investigate the uranium concentration in the calcrete ores on the edges of the salt lakes.

The group tried hard to buy an ex-Soviet nuclear weapon through their numerous contacts in Russia and spent millions of US dollars attempting to develop laser technology in Japan to enrich uranium, presumably with the aim of fabricating a nuclear explosive. It also tried to develop effective biological weapons using anthrax, botulinum toxin and the Ebola virus. The group also recruited a number of computer programmers who were installing computer systems in many of the top Japanese corporations. There is some concern that other things, such as remote transmitters or eavesdropping devices, may have been installed along with legitimate software.

The AUM group involved itself in a disturbingly wide range of activities and was the first terrorist group to use a weapon of mass destruction. It made nine attempts to use chemical or biological weapons. Finally, the sarin attack in Tokyo was successful.

There is a serious risk that terrorist groups will acquire and use WMDs. It is, therefore, crucial that countries develop effective methods of countering such a form of terrorism and implement them with a sense of urgency.

9 What can counter-terrorism do?

Analysts who study trends in international terrorism, including those in the main intelligence and security agencies, generally agree that for the foreseeable future terrorism will continue to be a serious threat and that there is a grave risk that terrorist violence may escalate to the use of weapons of mass destruction.

An obvious measure to reduce the risk of terrorism with a WMD is to prevent terrorists acquiring the materials essential for the fabrication of such a weapon. Nevertheless, security on these materials is often very lax; applying effective protection measures is difficult in a democracy. Consequently, counterterrorism activities by the police and intelligence agencies have the most important role to play in counterterrorism measures. In the words of the Report of the National Commission on Terrorism, US Congress, entitled *Countering the Changing Threat of International Terrorism*: "Good intelligence is the best weapon against international terrorism." Good intelligence requires cooperation between intelligence agencies, both within regions and internationally. The negotiation of appropriate regional and international agreements, and strict adherence to

> "Good intelligence is the best weapon against international terrorism"

them, are crucial. Legal sanctions must be written into these agreements to reduce the risk of violations.

Protecting key materials

If a terrorist group wants to fabricate a WMD it must acquire, legally or illegally, one or more of a number of key materials. To make nerve agents the terrorist group would need ingredients such as phosphoryl chloride and dimethylamine; for biological weapons it would need, for example, access to anthrax, plague, or botulinum bacteria; for nuclear weapons it would have to have plutonium or highly enriched uranium; and for radiological weapons it would need significant amounts of a radioactive isotope, such as cobalt-60 or caesium-137.

Clearly, action to prevent the acquisition by terrorists of weapons of mass destruction should focus on the physical protection of the key materials, which must take into account the relatively small amounts of the materials needed to make a WMD.

Society may decide that the risk of terrorists acquiring and using a weapon of mass destruction, and the awesome consequences of such use, are sufficiently serious that some activities should be given up. An obvious example is the reprocessing of spent nuclear-power reactor fuel to separate the plutonium from it. When the plutonium is in spent nuclear fuel elements it is self-protecting because it is so radioactive that any person approaching it would soon die. When it is separated from them it could be acquired and used by terrorists to fabricate a primitive nuclear explosive.

Nuclear smuggling

Counterterrorist agencies are particularly concerned that terrorists will acquire fissile materials, plutonium and highly enriched uranium, from nuclear smugglers and use them to fabricate a nuclear explosive. There is so much fissile material, suitable for use in nuclear weapons, in the world that could be used to make nuclear weapons that there are likely to be opportunities for smugglers to get hold of some.

These materials are ideal for smuggling. A kilogram of weapon-usable plutonium, for example, would probably be worth 1–2 million US dollars on the black market. About the size of a golf ball, it could be smuggled across borders very easily.

A problem for those combating the smuggling of weapon-usable plutonium or highly enriched uranium is that the materials are difficult to detect, in airports or at borders for instance, using radiation detectors such as Geiger or scintillation counters. The materials are very weakly radioactive. With some shielding (using lead, for example), very little radiation would escape from the shielding so that a very sensitive system for detecting the small amount of escaped radiation would be needed.

About the size of a golf ball, weapon-usable plutonium could be smuggled across borders very easily

New equipment to detect smuggled fissile materials has very recently been developed. The system uses X-rays that can penetrate shielding materials and, for example, the walls of cargo containers. If the X-rays react with plutonium or uranium nuclei they induce fission reactions. The neutrons emitted in the processes pass through the shielding material and can be detected by neutron detectors operated by security personnel and customs officers.

Smart smugglers, however, are likely to frustrate the new system by surrounding the plutonium or highly enriched uranium with a substance with a high hydrogen content, such as a plastic or paraffin wax, which would absorb the neutrons so that they are not recorded by the neutron counters.

The republics of the former Soviet Union are a potential source of illegal nuclear materials. In the words of Bill Clinton:

The breakup of the Soviet Union left nuclear materials scattered throughout the newly independent states and increased the potential for the theft of those materials, and for organized criminals to enter the nuclear smuggling business. As horrible as the tragedies in Oklahoma City and the World Trade Center were, imagine the destruction that could have resulted had a small-scale nuclear device exploded there.

More than 100 scientific and industrial institutions and facilities in Russia and some other ex-Soviet republics keep nuclear fissile materials. Many of these establishments are in cities virtually controlled by the Russian Mafia. Estimates suggest that there are about 6,000 Mafia gangs in Russia with a membership of more than 100,000 criminals. It is hardly surprising that many fear that nuclear material is in the hands of the Mafia or will be acquired by it.

There are even fears that some ex-Soviet nuclear weapons are missing. In the mid-1980s, when the former Soviet Union collapsed, there were about 30,000 nuclear weapons on Soviet soil. Will these nuclear weapons in the ex-Soviet arsenal be kept securely? The majority of weapons may be relatively secure while they are in the hands of the military and the security service. But

Weapon production and testing sites in Russia

the risk that a few of them will be illegally acquired is significant. The relatively small tactical nuclear weapons, such as nuclear artillery shells or nuclear land mines, are the most vulnerable to theft. The trunk of a car would be sufficient transport for one of these.

If any do get into the wrong hands, we probably will not know. It is, to say the least, very doubtful that a complete inventory of the ex-Soviet nuclear weapons exists. The Soviet bureaucracies were so confident that their nuclear weapons were safe, being so closely guarded by the KGB and the military, that they probably did not bother to record them all.

there are about 6,000 Mafia gangs in Russia with a membership of more than 100,000 criminals

Fears that a flourishing black market exists, involving the smuggling of fissile materials from Russia and other ex-Soviet republics, have been increased by a number of incidents. For example, the authorities in Prague seized 3 kilograms of uranium, enriched to 87 percent in the isotope uranium-235, in December 1994. And, on August 10, 1994, about 330 grams of weapons-grade plutonium and 1 kilogram of lithium-6 (an isotope used to produce lithium-6 deuteride for use in thermonuclear weapons) were captured in Munich. A Colombian and two Spaniards had been persuaded by German secret agents to bring the material to Germany.

On December 14, 1994, the Czech authorities seized 3 kilograms of highly enriched uranium in Prague and three men were arrested. One was a Czech nuclear physicist, which indicated that nuclear scientists may be becoming involved in nuclear smuggling. This would be a serious development, as they would of course be expert in identifying which materials to steal and how to handle them. These seizures were usually made after the authorities were tipped off, although in some cases intelligence agents had penetrated smuggling rings. The known incidents are probably the tip of an iceberg.

Smugglers are perhaps more likely to use routes to smuggle

materials out of Russia through Eastern Europe than through Germany, as these other routes are likely to be less well policed. There are, for instance, well-established smuggling routes from Russia through the Baltic states, Lithuania, Latvia, and Estonia. Another is through the southern ex-Soviet republics into Turkey and then on into the Middle East. Yet another likely route out of Russia is through Vladivostok to, perhaps, China. Given these, and other, possible routes of smuggling materials, it is impossible to know the extent of the activity.

Nuclear smuggling is one of the links between organized crime and terrorism. Ian O. Lesser, of the RAND organization in California, explains that

> the enormous sums of money involved, as well as numerous points of contact between leading mafias and legitimate institutions, can facilitate acts that would be difficult for politically motivated groups to undertake—and pay for—on their own. This is a particular risk in relation to nuclear terrorism. Although details remain murky, Russian mafias are already reported to be involved in obtaining and smuggling nuclear materials, and in the most extreme case, perhaps even small nuclear weapons (such as nuclear land mines).

The control of nuclear smuggling, like other types of smuggling, is an extremely difficult task, requiring top-rate intelligence. As Matthew Bunn, a Harvard expert on nuclear smuggling, argues: "once nuclear materials are removed from the enterprise, much of the battle is already lost. Finding stolen material within a country or detecting and interdicting its passage across borders are Herculean tasks, in most cases only practicable if good intelligence and police work tell officials where to look." The former Soviet republics urgently need financial help and expertise to effectively monitor and control the illicit trade in nuclear materials.

The main stocks of ex-Soviet nuclear materials remain in Russia, but the risk of their being smuggled out is considerable—mainly because of the current state of political turmoil and

Russia's poor and deteriorating economy. But perhaps the most serious factor is the lack of a culture of nuclear safety and security in Russia.

Such a culture has never evolved and there is, therefore, little national or regional regulation of nuclear activities, including the storage of nuclear materials. Moreover, the poor morale of employees at Russian nuclear institutions and facilities, continually worsened by poor and often delayed pay, seriously increases the chance of the theft and sale of highly valuable nuclear materials. In 1999, the average salary of the workforce in Russia's nuclear facilities was about 45 US dollars per month.

One indication of a lack of a nuclear security culture is the lax methods of the disposing of radioactive waste in Russia. Investigations by the Norwegian Environmental Foundation Bellona have shown that Russian nuclear submarines routinely discharge radioactive liquids into the oceans; that reactors from decommissioned nuclear-powered submarines are simply dumped into the oceans, as is other radioactive waste; and that huge areas around nuclear establishments, civil and military, have been severely radioactively contaminated.

In 1999, the average salary of the workforce in Russia's nuclear facilities was about 45 US dollars per month

People who have visited nuclear facilities in the former Soviet Union have described how these have deteriorated: "holes in perimeter fences, non-functioning alarm systems, and paper records that fail to match physical inventories of materials." Under these conditions, nuclear materials can easily go astray.

In the short term, international help is needed to improve nuclear security by training relevant staff and to assist the authorities to develop and adopt adequate nuclear regulatory oversight. Russia's nuclear regulating body—Gosatomnadzor (GAN), created in December 1991—is virtually powerless. Unless GAN is given the legal power to improve nuclear safety and security, lax practices are bound to continue.

The evolution of a culture of nuclear security and safety and the establishment of an adequate national regulatory organization with the power to enforce action against reactor operators and other nuclear establishments will not be easy in a state described by Jonathan Steele as "a country without law or a sense of social responsibility among the elite."

If and when a culture of nuclear security and safety has been established, the Russians will need financial help among other things, to prevent the theft and smuggling of nuclear materials, and the total amount of money involved will be large. Given the current economic difficulties in the G-7 countries and the economic, social, and political turmoil in Russia, the chances of significant amounts of money being provided are small.

Regional and international agreements

Regional measures

Terrorism does not recognize national boundaries: to be effective, counterterrorist police and intelligence agencies need to act across borders. Although there are a number of bilateral arrangements between countries to achieve cooperation between national police forces and intelligence agencies, there are very few regional arrangements. One regional agreement, establishing the Europol, is, however, in operation.

The idea of Europol, based in The Hague, originated in the 1992 Maastricht Treaty on European Union to improve the effectiveness of cross-border police cooperation and the sharing of intelligence to combat terrorism, drug trafficking, and other serious forms of international crime affecting two or more Member States of the European Union (EU). Europol has been operating unofficially since 1994 but the Europol Convention, which established it, only came into operation on October 1, 1998. The EU Member States signed the Europol Convention in July 1995 and by June 15, 1998 it had been ratified by all of them.

The Convention requires Europol "to improve the effectiveness of the competent authorities in the Member States and cooperation between them in an increasing number of areas." These include preventing and combating: terrorism; unlawful drug-trafficking; trafficking in human beings; crimes involving clandestine immigration networks; illicit trafficking in radioactive and nuclear substances; illicit vehicle trafficking; combating the counterfeiting of the euro; and money-laundering associated with international criminal activities.

The Convention requires Europol

> to facilitate the exchange of information between Member States; to obtain, collate and analyse information and intelligence; to notify the competent authorities of the Member States without delay of information concerning them and of any connections identified between criminal offences; to aid investigations in the Member States; and to maintain a computerized system of collected information.

Europol has experienced some teething problems and the authorities in some EU countries, including the Netherlands, have been critical of it. But Emanuel Marotta, the Deputy Director of Europol, points out that:

> Work in the field of illegal drug trafficking and other specific areas of organized crime has borne some outstanding successes. A network of 45 liaison officers representing the major law enforcement agencies of the Member States is already in place, and furthermore, Europol already has a pool of a dozen analysts working in close cooperation with these liaison officers.

Nearly 200 people are employed at the Europol headquarters in The Hague; Europol's annual budget is 35 million euros. Since mid-1999, it has been promoting international cooperation against terrorism including the exchange of information about terrorist crime among the EU Member States via the network of liaison officers and maintaining a databank about terrorists and their activities. Clearly, this is only the beginning; the role of Europol will increase substantially in the future.

On September 21, 2001, ten days after the terrorist attacks on New York and Washington, the EU Interior Ministers agreed to give Europol new powers, including: a European arrest warrant (extradition would be replaced with a procedure for handing over perpetrators of terrorist attacks on the basis of such a warrant); coordination of civil protection measures; close cooperation between police and intelligence services; close cooperation between EU and US security agencies; better access to data; rapid transfer of relevant information to Europol; a team of counter-terrorist specialists at Europol; stricter procedures in connection with issuing visas; and improvement of airport security and aviation safety standards.

Europol is now likely to expand its work by cooperating with non-European Union countries. Time will tell how successful it is in improving international cooperation in combating terrorism. If it is successful it is likely to be a model for the establishment of other regional counterterrorism agencies.

International measures

At the international level there are twelve major multilateral conventions and protocols, dating back to 1963, related to states' responsibilities for combating terrorism. But many states are not yet party to these legal instruments, or are not yet implementing them. There are now a number of important UN Security Council and General Assembly Resolutions on international terrorism, dealing with specific incidents.

The major terrorism conventions and protocols are on: the safety of aviation; aircraft hijackings; acts of aviation sabotage such as bombings aboard aircraft in flight; attacks on senior government officials and diplomats; the taking of hostages; the unlawful taking and use of nuclear material; unlawful acts of violence at airports; terrorist activities on ships; terrorist activities on fixed offshore platforms; and chemical marking to facilitate detection of plastic explosives, e.g., to combat aircraft sabotage.

UN General Assembly Resolutions include the International Convention for the Suppression of Terrorist Bombing and the International Convention for the Suppression of the Financing of Terrorism.

The United Nations Office for Drug Control and Crime Prevention (ODCCP) is a global leader in the fight against illicit drugs and international crime, including international terrorism. Established in 1997, ODCCP consists of the United Nations International Drug Control Programme (UNDCP) and the United Nations Centre for International Crime Prevention (CICP).

ODCCP has approximately 350 staff members worldwide. Its headquarters are in Vienna and it has twenty-two Field Offices, as well as Liaison Offices in New York and Brussels.

Also established in 1997, the Centre for International Crime Prevention (CICP) is the United Nations office responsible for crime prevention, criminal justice and criminal law reform. The CICP works with Member States to strengthen the rule of law, promote stable and viable criminal justice systems and combat the growing threat of transnational organized crime through its Global Programme Against Corruption, Global Programme Against Organized Crime, Global Programme Against Trafficking in Human Beings, and its Terrorism Prevention Branch (TPB).

The urgent need for effective intelligence

Considering the number of successful terrorist attacks that have taken place over the years, it must be said that the security and intelligence agencies have not risen to the challenge presented by terrorist groups. British experience in combating both nationalist and loyalist terrorism in Northern Ireland drives this point home. A large number of intelligence agencies have operated in Northern Ireland, working for the Royal Ulster Constabulary, Special Branch and military intelligence. There is considerable rivalry between these agencies, which causes a significant loss of efficiency. Moreover, there is little cooperation and sharing of information between them. Each agency has its own leadership. The Northern Ireland situation is indicative of a general problem.

Each intelligence agency has its own leadership

Some improvements have been made in the way the intelligence agencies operate. In the words of Ely Karmon:

> From the organizational point of view, the security and intelligence agencies have taken serious steps to improve their capabilities. The FBI has tripled its counterterrorism force since the World Trade Center attack and the CIA has created a Counter-Terrorism Center (CTC) to deal with the threat at the highest civilian and military level. The German authorities have greatly enhanced the police and security units dealing—successfully at that—with right-wing activities.

Nevertheless, the recent terrorist attacks in New York and Washington bring home the fact that the shortcomings of intelligence agencies far outweigh their successes. The intelligence and security communities should take the threat of the acquisition and use of weapons of mass destruction much more seriously than they do at the moment.

Monitoring the communications of terrorist groups—the activity known as signal intelligence (SIGINT)—has been crucial to this end. Modern terrorists can, however, take steps to protect their communication systems, including the use of the most up-to-date methods of encryption which are exceedingly difficult to counter.

The penetration of terrorist groups by undercover intelligence agents or double agents (human intelligence or HUMINT) is, therefore, of critical importance. In fact, counterterrorism is likely to succeed only if HUMINT can be made effective. This is why it is, to say the least, not going to be easy to defeat terrorism, particularly fundamentalist terrorism. The problem is that the fundamentalist groups are by far the most difficult to infiltrate.

Nevertheless, it is known that the intelligence agencies have prevented some attacks by international terrorist groups but their successes are not generally publicized. In one known instance, the infiltration of agents into the group involved in the 1993 bombing of the World Trade Center foiled a plan to bomb the United Nations building and the Lincoln tunnel in New York. Planned attacks by the Al Qaeda organization and its allies in, among other places, Israel, Jordan, Pakistan, some Western European countries, and the USA have been foiled, leading to the arrest of many of the terrorists involved.

Experience shows that setting up effective intelligence activities against terrorist groups is extremely challenging. Rivalries between intelligence agencies within countries and lack of cooperation in intelligence matters between countries seriously reduce the effectiveness of intelligence, to say the least.

The following measures are urgently needed:

- One influential and powerful person, who has adequate access to the political leadership, should lead intelligence and security agencies within countries.
- The leaders of national intelligence agencies should be in frequent contact.
- National databanks should be integrated and made available to regional and international authorities.
- Effective and single leadership, international cooperation and flexibility should be the keys to good counterterrorism intelligence. In the context of improving international cooperation, an encouraging development is the establishment of the Terrorism Prevention Branch of the United Nations, within the Centre for International Crime Prevention.
- The monitoring and control of the trade, within and between states, in the materials needed by terrorists to fabricate chemical, biological and nuclear weapons should be considerably improved.

• Current international agreements, conventions, and treaties relating to chemical, biological, and nuclear weapons should be strengthened, particularly by improving existing safeguard measures and adding improved ones, and by introducing legally enforceable sanctions if violations occur.

In sum, the intelligence and security agencies, in their fight against terrorism, face an awesome task that will require the acquisition of any new technological developments relevant to counter-terrorist activities, a close study of new terrorist threats, and, perhaps most importantly, an imaginative approach to the issues.

In the age of the Internet, knowledge is available to all

In the age of the Internet, knowledge is available to all. This, and the revolution in communications, have had a considerable impact on society and have removed one of the advantages of the intelligence community. In future, success in countering terrorism will depend on the effective application of ingenuity and innovation.

10 What does the future hold?

The future proliferation of ballistic missiles

The spread of WMDs to countries that do not currently have them must be expected to continue, as must the proliferation of ballistic missiles. The two are closely related: ballistic missiles are ideal systems for delivering WMDs, particularly over long distances. Ballistic missiles are expensive to develop and to buy, and they carry relatively small payloads. It is, therefore, hardly worth acquiring them just to deliver conventional warheads—bluntly put, they don't do enough damage to justify the expense.

conventional warheads don't do enough damage to justify the expense

Ballistic missiles are divided, rather arbitrarily, into short-range, medium-range, long-range, and intercontinental. Short-range are those with ranges of less than 150 kilometers; medium-range have ranges between 150 and 1,000 kilometers; long-range have ranges between 1,000 and 5,000 kilometers; and intercontinental are those with ranges greater than 5,000 kilometers. They are also divided into tactical and strategic ballistic missiles. Tactical ones are generally those that can deliver warheads of relatively small mass, a few hundred kilograms, over

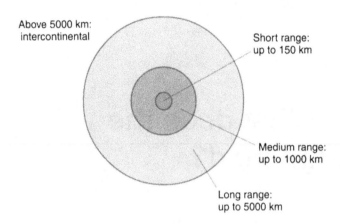

Missile ranges

ranges of about 1,000 kilometers. Strategic ballistic missiles are those with longer ranges and able to deliver heavier payloads.

Thirty-six countries possess ballistic missiles of some type. Twenty-one of them have deployed short-range ballistic missiles; twenty-four have deployed medium-range ballistic missiles; twelve have long-range missiles; and five have intercontinental ballistic missiles. Fourteen countries are producing ballistic missiles; three of these (China, North Korea and the USA) export them as part of the global arms trade. In addition, several Third World countries (Brazil, Libya, Serbia, and South Africa) are developing or are capable of developing ballistic missiles indigenously. About twenty Third World countries have ballistic missiles in their arsenals. Given the number of countries with ballistic missiles and their increasing availability in the global arms trade, terrorist groups are eventually likely to get their hands on some and use them to deliver WMDs.

All countries with nuclear weapons—China, France, India, Israel, Pakistan, Russia, the UK, and the USA—have deployed them on ballistic missiles. The countries thought to be developing them—among these are Iran, Iraq, and North Korea—already possess ballistic missiles, which they could thus decide to arm with nuclear warheads.

Many of the countries suspected of deploying biological and chemical weapons could deliver them by ballistic missiles. Because ballistic missiles are readily available, all sorts of terrible weapons can be easily delivered by those countries that have them. The future proliferation of ballistic missiles, and their use by terrorists to deliver WMDs, are serious threats to security.

Nuclear terrorism

In addition to stealing or otherwise acquiring fissile material and fabricating and detonating a primitive nuclear explosive, there are a number of other nuclear activities in which a terrorist group may become involved: attacking a nuclear-power station reactor to spread radioactivity far and wide; attacking the high-level radioactive waste tanks at reprocessing plants to spread the radioactivity in them; attacking a plutonium store to spread the plutonium in

the risk of nuclear terrorism is considerable

it; attacking, sabotaging, or hijacking a transporter of nuclear weapons or nuclear materials; and making and detonating a radiological weapon, a so-called "dirty bomb," to spread radioactive material.

Apart from the use of a dirty bomb, which could cause much disruption but relatively few fatalities (see pages 38–9), all of these types of nuclear terrorism have the potential to cause large numbers of deaths. Given the potentially serious consequences of a nuclear terrorist attack, policy makers have to make the difficult judgement about the probability of its happening.

But they must be aware that the risk of nuclear terrorism is considerable. Mohamed El Baradei, the Director General of the International Atomic Energy Agency, warns:

The proliferation of ballistic missile capabilities

Japan
Advanced Space Launch Program (Cap)

N. Korea (Export)

S. Korea

Taiwan

China (export)

Russia

India

Pakistan

Iran

Iraq

Egypt

Israel

France

Spain
Space Launch Program (Cap)

Libya
Al Fattah SRBM Program continuing (Dev)

Serbia
Tested the K-15 Krajina SRBM with a 150 km range. Announced in November 1994 that it was developing the 400 km-range modified Scud SRBM (Dev)

S. Africa
Instigated Arniston MRBM Program with Israeli assistance in 1980s. The program has been officially terminated (Cap)

Argentina

Brazil
Developing Space Launch Capability Missile Program "terminated" in late 1980s (Cap)

USA (export)

Countries producing and/or exporting ballistic missiles

Countries developing (Dev) or with the capability to develop (Cap) ballistic missiles

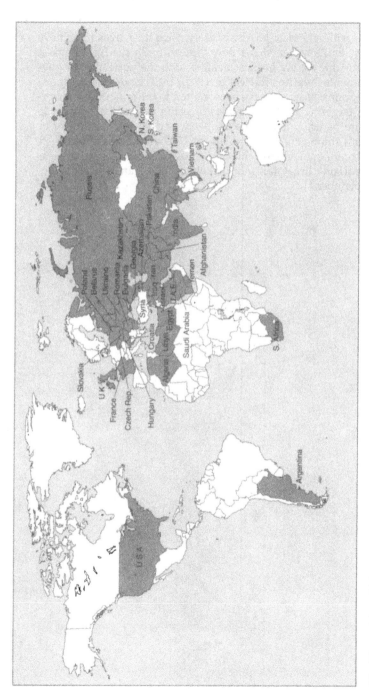

The 36 countries which possess ballistic missiles + countries making, exporting, and capable of developing ballistic missiles

The willingness of terrorists to sacrifice their lives to achieve their evil aims creates a new dimension in the fight against terrorism. We are not just dealing with the possibility of governments diverting nuclear materials into clandestine weapons programs. Now we have been alerted to the potential of terrorists targeting nuclear facilities or using radioactive sources to incite panic, contaminate property, and even cause injury or death among civilian populations. The willingness of terrorists to commit suicide to achieve their evil aims makes the nuclear terrorism threat far more likely today than it was before September 11.

Terrorist attack on a nuclear-power station

Instead of fabricating and exploding a nuclear weapon, a terrorist group may decide to attack a nuclear facility. A group with significant resources could attack and damage nuclear-power plants. There is disagreement, however, about how much damage would be done and how many people harmed by such an attack. It is probably true that attacks on nuclear-power plants that could do a great deal of damage and cause many fatalities have a relatively small chance of success. But the damage caused and the number of people killed by a successful terrorist attack on a nuclear-power plant could be so catastrophic that even a small risk of such an attack is not acceptable.

In a nuclear-power station there are two potential targets for a terrorist attack: the reactor itself and the ponds storing the highly radioactive reactor fuel. An attack on the reactor could cause the core to melt down, as happened during the 1986 accident at the Chernobyl reactor, or cause a loss of the coolant, usually water, that removes heat from the core of the reactor, as happened during the accident at Three Mile Island in 1979.

Spent fuel elements are normally kept in storage ponds for five or ten years under about 3 meters of water before they are finally either disposed of in a geological repository or sent to a reprocessing plant where the plutonium automatically produced in the fuel elements is chemically separated from unused uranium and fission products in the fuel elements. The ponds are normally built close to the reactor building. The buildings containing the spent fuel ponds are less well protected than the reactor and are, therefore, more attractive targets than the reactor building.

Terrorists could target a reactor or spent fuel pond by: using a truck carrying high explosives and detonating it near a critical part of the target; exploding high explosives carried in a light aircraft near a critical part of the target; crashing a hijacked commercial airliner into the reactor building or spent-fuel pond; attacking the power station with small arms, artillery, or missiles and occupying it; or attacking the power lines carrying

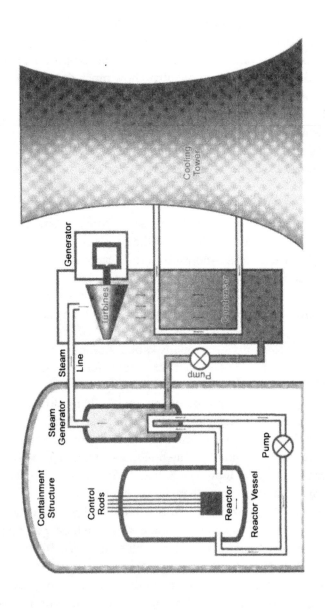

The components of a nuclear power plant

electricity into the plant. Alternatively, a terrorist group may infiltrate the plant so that some of its members, or sympathizers, can sabotage it from inside. A saboteur may attack the systems cooling the reactor core, or drain water from the cooling pond. This could cause the temperature of the reactor core to rise, resulting in a release of radioactivity from the core, or cause the temperature of the spent fuel rods to rise, again resulting in a release of radioactivity.

Potential consequences of a terrorist attack on Sellafield

The high-level liquid waste tanks

It is hard to think of a nuclear terrorist attack which could, at least in theory, be more catastrophic than a successful attack on either the tanks at the nuclear site at Sellafield, England or those at La Hague, France, that contain the liquid fission products separated from spent reactor fuel elements by the two reprocessing plants; or on the stores holding the plutonium separated by the reprocessing plants.

If an aircraft on one of the commercial flight paths taking it closest to Sellafield were hijacked and crashed into the nuclear site, it would take only about four or five minutes to reach its target. This time is too short for the authorities to detect, intercept, interrogate and take remedial action to prevent the attack.

Nuclear facilities are vulnerable to terrorist attacks. Particularly so would be an attack in which a large commercial aircraft, such as a Boeing 747 carrying a full load of fuel, is dived from a high altitude into the liquid high-level waste (HLW) tanks or the plutonium store at Sellafield. A fully laden jumbo jet traveling at between 200 and 300 meters a second would have a very large momentum and the crash would have a huge impact. In addition, the aircraft may be carrying about 150 tons of aviation fuel and the crash would create a ferocious fire.

Highly radioactive liquid waste, arising from the operations of the two reprocessing plants at Sellafield, is stored in twenty-one cooled tanks. Fourteen of these tanks are kept full; the other seven are kept empty so that liquid waste can be pumped into them in an emergency. Without cooling, the heat produced by the radioactive decay of the isotopes in the liquid waste would cause the liquid to boil and the tanks to explode. The total volume of HLW is limited to a maximum of 1,575 cubic meters. The volume of HLW in the tanks is currently at this level.

Highly radioactive liquid waste, at Sellafield, is stored in twenty-one cooled tanks

So far as the contamination of the human environment and damage to human health are concerned, the most important radioisotope in the HLW tanks at Sellafield is caesium-137, a particularly hazardous radioisotope to people exposed to it.

Any of the liquid tanks at Sellafield that survived the initial impact of the terrorist attack considered here are likely to dry out because the impact will cut off the cooling system. Caesium-137 is volatile and the bulk of it will escape into the atmosphere over, say, a two-day period. It would not be possible to establish emergency cooling for weeks because of the high level of radioactivity in the area.

In the first minute or so after the accident, the fire caused by burning aviation fuel is likely to produce a fireball rising to an altitude of up to between 1 and 2 kilometers. After the first minute or so, radioactivity will continue to be released but will not rise more than a few meters into the atmosphere.

The total amount of caesium-137 in the tanks weighs about 1,980 kilograms. It is instructive to compare the radioactive contamination potentially caused by a terrorist attack on Sellafield with that caused by the reactor accident at Chernobyl on April 26, 1986. The Chernobyl accident released about 25 kilograms of caesium-137. Each of the HLW tanks contains over 3.5 times as much caesium-137 as was released at Chernobyl.

Figures given by the United Nations Scientific Committee on the Effects of Atomic Radiation (UNSCEAR), suggest that the number of fatal cancers produced by the Chernobyl accident is 30,000. If all the caesium-137 in just one tank at Sellafield were released, the radioactivity could cause about 170,000 fatal cancers. Depending on the strength and direction of the winds at the time of the release of the radioactivity, these deaths will occur in the United Kingdom, Ireland, and parts of Europe and perhaps even further afield.

Figures suggest that the number of fatal cancers produced by the Chernobyl accident is 30,000

If the terrorist attack on the HLW tanks releases more radioactivity than is assumed above then the number of fatal cancers will be proportionally larger. In the unlikely case that all the caesium-137 in the tanks is released, the number of people suffering fatal cancers as a result could reach a total of about 2.25 million.

A terrorist attack on the plutonium stores at Sellafield

The two plutonium stores at Sellafield contain the plutonium separated from spent nuclear-power reactor fuel elements in the two reprocessing plants. Currently, the stores contain about 71 tons of plutonium in the form of plutonium dioxide. A terrorist attack on the plutonium stores could contaminate the environment. If plutonium is inhaled or ingested, the radiation given off by it can be particularly damaging to the cells of the body. When outside the body, plutonium does not present a significant hazard, but it is particularly toxic when inhaled into the lungs.

The main task after a release of plutonium into the human environment is the evacuation and decontamination of land contaminated. Particles of plutonium that have fallen to the ground are still a potential health hazard. If the particles are disturbed, or blown by the wind, they can become airborne again.

The concentration of resuspended plutonium particles will be much less than the original concentration of plutonium particles in the cloud, but they will remain a health hazard until the area is decontaminated, a very time-consuming operation.

The level of land contamination with plutonium isotopes that would require decontamination (by, for example, the removal of topsoil) depends on the circumstances. The UK National Radiological Protection Board requires land contaminated by more than a given level to be evacuated and decontaminated. If evenly distributed, a kilogram of plutonium in the Sellafield store would, on average, contaminate more than 300 square kilometers to the level at which the National Radiological Protection Board (NRPB) recommends evacuation. A terrorist attack on the plutonium store at Sellafield could seriously contaminate a huge area of land.

Biological and chemical weapons are easier for terrorists to develop and produce than nuclear ones. And genetic engineering may make biological agents very attractive for terrorists. Nevertheless, there are good reasons to believe that future fundamentalist terrorists are more likely to use nuclear explosives than biological or chemical agents.

A nuclear explosion fits well with fundamentalist apocalyptic vision, because of, among other things, media "appeal." And as Melissa Chirico explains:

> first, terrorist groups are often bound to visuals. A nuclear blast provides immediate and extensive destruction. Second, a credible nuclear terrorist threat would have enormous coercive power because of the public's historical fear of nuclear explosions. As the Three Mile Island and Chernobyl incidents have shown, a nuclear blast is likely to result in mass panic and hysteria including severe anxiety, post-traumatic stress, epidemics, and paralysis of human life. Another proposed motivation for terrorist groups to use nuclear weapons has stemmed from the observation that they often attempt to imitate governments to make themselves appear more legitimate. By using the weapons of powerful states, terrorist groups may hope to be treated like powerful states as well.

But the situation is by no means clear-cut. There is no way of predicting what extraordinary weapons military geneticists may develop.

Genetic engineering

Genetic engineering has given scientists the capability to produce new biological warfare agents. The findings from the Human Genome Project and the Human Genome Diversity Project are likely to be used to develop new biological weapons. Man-made biological agents may be much more suitable for use in biological weapons than natural ones. They may be better able to withstand changes in environmental factors such as humidity and temperature, increasing their chance of surviving storage and dispersal.

The genes that determine the lethality of the bacteria that produce diseases, like anthrax and plague, can be identified. These genes can easily be sliced into bacteria that are normally harmless; an undergraduate biologist could do so. Deadly genes from anthrax, for example, could be put into the bacteria *Escherichia coli*, very prolific bacteria in the human gut.

The newly made deadly *E. coli* could be rapidly produced in large quantities. They would be particularly lethal because the body would be familiar with them and so would be unlikely to produce antibodies. People infected with them would, therefore, not have the ability to fight the disease. Some states and a terrorist group have already produced genetically engineered biological agents. The AUM group successfully re-engineered *E. coli*, placing botulinum toxin within it.

Genetic engineering to produce more sophisticated biological-warfare agents has proved successful. In 1998, A. P. Pomerantsev and his fellow researchers at Russia's State Research Center for Applied Microbiology in Obolensk described how they used genetic engineering techniques to insert deadly genes from a harmless bacteria *Bacillus cereus* into the anthrax bacteria *Bacillus anthracis.* The strain used in this work was bred to be resistant to antibiotics. The product was an entirely new form of anthrax resistant to vaccines. Colonel Arthur Friedlander, an American biological-warfare expert, explained that the new organism is likely to cause: "disease by a different mechanism than that used by naturally occurring anthrax strains." Future biological-warfare agents, for use against humans, animals and crops, will be based on genetically engineered organisms: these could be extremely contagious; consistent in their effects; relatively safe to handle; stable under production, while in storage, in munitions, and during transportation; difficult for the enemy to identify; and impossible for the enemy to vaccinate against. Tomorrow's biological-warfare agents are likely to be of great interest both to the military and to terrorists.

> *Genetic engineering to produce more sophisticated biological-warfare agents has proved successful*

Biowar against ethnic groups

Another possible type of genetic weapon involves "genetic-homing" weapons, such as specific genetic viruses, that could "target a genetic structure shared by particular ethnic groups or specific human attributes." A recent publication by the Stockholm Peace Research Institute explains that genetic differences between ethnic groups "may in many cases be sufficiently large and stable so as to possibly be exploited by using naturally occurring, selective agents or by genetically engineering organisms and toxins."

Using differences in DNA from different groups of people, it may be possible to attach elements to DNA to kill people from a specific group. Differences between ethnic groups may be determined and then exploited in this way. There are genetic differences between, for example, those with blacker or whiter skin in the control of pigment production and distribution. These differences may not be absolute because there is usually no sharp dividing line between groups of people, but those developing such weapons in the future are unlikely to be worried about damaging some people outside the target group.

This horrifying form of ethnic cleansing is not science fiction. In 1996, Vivienne Nathanson, head of science and ethics at the British Medical Association, warned the World Medical Congress in South Africa that: "One could imagine in Rwanda, a weapon which targeted one of the two tribal groups, the Tutsi and Hutu. While these weapons do not, as far as we know, exist, it is not far away scientifically." And, in 1996, General Bo Rybeck, the former head of Sweden's Defense Research Establishment, told a meeting of the International Committee of the Red Cross that genetic weapons may be "just around the corner."

Future terrorists may well use genetic engineering to produce biological weapons to attack a specific ethnic group, with devastating effects

All industrialized countries, and some developing ones, like India and China, have biotech industries and most are active in genetic engineering. The more numerous the countries that are involved in biotechnological developments, the more likely it is that terrorists will acquire lethal and effective biological agents. A large terrorist group could genetically engineer biological agents itself, as the AUM group has shown. It would need to employ biologists and genetic engineers. The capital equipment for such a program need be no more than about 30,000 US dollars; the running costs would also be about 30,000 US dollars a year. Future terrorists may well use genetic engineering to produce biological weapons to attack a specific ethnic group, with devastating effects.

Cyberterrorism

The term "cyberterrorism" links the virtual world of computers, which store, process, and communicate information, with the violent and unpredictable world of terrorism. Cyberterrorism must not be confused with other abuses of computers, such as computer crime, hacking, information warfare, computer tapping. Mark Pollitt, a special agent for the FBI, has given a suitably narrow definition of cyberterrorism: "Cyberterrorism is the premeditated, politically motivated attack against information, computer systems, computer programs, and data which result in violence against noncombatant targets by sub-national groups or clandestine agents."

> *Tomorrow's terrorist may be able to do more damage with a keyboard than with a bomb*

The risk that future terrorists will attack the computer networks in industrialized countries, producing massive economic loss and social disruption, is real and serious. The consequences of cyberterrorism are described in a book about computer security, *Computers at Risk*, produced by the American National Academy of Sciences:

> Increasingly, America depends on computers. They control power delivery, communications, aviation, and financial services. They are used to store vital information, from medical records to business plans to criminal records. Although we trust them, they are vulnerable—to the effects of poor design and insufficient quality control, to accident, and perhaps most alarmingly, to deliberate attack. The modern thief can steal more with a computer than with a gun. Tomorrow's terrorist may be able to do more damage with a keyboard than with a bomb.

Cyberterrorism has attractions for terrorists. It can be conducted at a great distance and costs little. It would not involve the use of explosives or weapons and, perhaps most importantly for the terrorist, it would attract a great deal of coverage in the media.

Computers and any problems with them fascinate the public and the media.

Concern about potential threats to the security of America's critical infrastructures led President Bill Clinton's White House to set up a Commission in 1996—the President's Commission on Critical Infrastructure Protection. The Commission examined the vulnerabilities to a wide range of threats, identifying eight infrastructures: telecommunications, banking and finance, electrical power, oil and gas distribution and storage, water supply, transportation, emergency services, and government services. In its report in October 1997 the Commission concluded that the threats to critical infrastructures were real and that, because these infrastructures are interconnected, they could be vulnerable in new ways. The report stated: "Intentional exploitation of these new vulnerabilities could have severe consequences for our economy, security, and way of life." The report went on:

> *Electrons don't stop to show passports*

> In the past we have been protected from hostile attacks on the infrastructures by broad oceans and friendly neighbors. Today, the evolution of cyberthreats has changed the situation dramatically. In cyberspace, national borders are no longer relevant. Electrons don't stop to show passports. Potentially serious cyberattacks can be conceived and planned without detectable logistic preparation. They can be invisibly reconnoitered, clandestinely rehearsed, and then mounted in a matter of minutes or even seconds without revealing the identity and location of the attacker.

The report emphasized the importance of "developing approaches to protecting our infrastructures against cyberthreats before they materialize and produce major system damage." As a consequence of the Commission's work, a number of agencies were established, including the National Infrastructure Protection Center, the Critical Infrastructure Assurance Office, the National Infrastructure Assurance Council, and private sector Information

Sharing and Assessment Centers. To protect its computer networks, the Pentagon established a Joint Task Force, Computer Network Defense.

To illustrate the potential vulnerability of critical computer systems to cyberterrorism, Dorothy E. Denning describes an exercise, conducted in June 1997 by the American National Security Agency. The point was to examine the vulnerability of US military computers and some civilian infrastructures to a cyberattack. She explains:

> Two-man teams targeted specific pieces of the military infrastructure, including the US Pacific Command in Hawaii, which oversees 100,000 troops in Asia. One person played the role of the attacker, while another observed the activity to ensure that it was conducted as scripted. Using only readily available hacking tools that could easily be obtained from the Internet, the NSA hackers successfully gained privileged access on numerous systems. They concluded that the military infrastructure could be disrupted and possible troop deployments hindered.

Cyberattacks on infrastructures, for instance those controlling water supplies, transportation systems and emergency services, could result in deaths and injuries. Others could cause considerable economic damage by, for example, interfering with stock market activities in ways that could precipitate inflation or depression.

The ultimate terrorist may totally disrupt society simply by sitting at a computer keyboard. A rival group may kill huge numbers with a genetically modified biological agent. George W. Bush recognized the twin threats in a speech in June 2001: "Our United States and our allies ought to develop the capacity to address the true threats of the 21st century. The true threats are biological and information warfare."

Appendix

Countries with military expenditure over 5,000 million US dollars p.a. (2001)

Country	Military Expenditure 2001 (thousand million US$)	Percentage of total
USA	322.3	38.6
Russia	63.7	7.6
China	46.1	5.5
Japan	39.5	4.7
UK	34.7	4.2
France	32.9	3.9
Germany	26.9	3.2
Saudi Arabia	24.3	2.9
Italy	21.0	2.5
India	14.2	1.7
South Korea	11.2	1.3
Brazil	10.5	1.3
Taiwan	10.4	1.2
Israel	10.4	1.2
Canada	7.8	0.9
Turkey	7.2	0.9
Spain	6.9	0.8
Australia	6.8	0.8
Netherlands	6.3	0.7
Mexico	5.7	0.7
Greece	5.5	0.7
Kuwait	5.0	0.6
Total these countries	719.2	86.1
World total	835.3	100.0

Further reading

Nonspecialist books dealing with weapons of mass destruction and terrorism using these weapons include:

James Campbell, *Weapons of Mass Destruction and Terrorism*, Inter-Pact, 1997.

Paul Leventhal and Yonah Alexander (eds.), *Preventing Nuclear Terrorism*, Lexington Books, 1987.

Adrian Dwyer, John Eldridge, Mike Kernan, *Jane's Chem-Bio Handbook*, Jane's Information Group, 2002.

Simon M. Whitby, *Biological Warfare against Crops*, Palgrave, 2002.

Wendy Barnaby, *The Plague Makers*, Vision Paperbacks, 1999.

Malcolm R. Dando, *Preventing Biological Warfare: The Failure of American Leadership*, Palgrave, 2002.

Ken Alibeck, *Biohazard*, Hutchinson, 1999.

Paul Wilkinson, *Terrorism and the Liberal State*, Macmillan, 1986.

US National Academy, *Computers at Risk*, National Academy Press, 1991.

Robert Hutchinson, *Weapons of Mass Destruction*, Weidenfeld and Nicolson, 2003.

Websites on weapons of mass destruction and terrorism

A selection of websites dealing with international security issues, including weapons of mass destruction and terrorism, and which have good links with similar organizations, follows:

http://irgs.humanities.curtin.edu.au/irgs/links/29.xml?method=show

http://www.acronym.org.uk/links.htm

http://www.basicint.org/nuclear/links.htm

http://www.mcis.soton.ac.uk/resources_links.htm

http://www.oxfordresearchgroup.org.uk

http://www.iiss.org

http://sipri.org

Websites with extensive sections on weapons of mass destruction and terrorism are:

www.nci.org

WWW.fas.org

www.globalsecurity.org

www.state.gov/t/np/wmd

Source notes

1 *National Strategy to Combat Weapons of Mass Destruction*, The White House, Washington, D.C., December 2002.

2 The Henry Stimson diaries are online at: www.doug-long.com/index.htm

3 S. Elworthy and P. Rogers, *The "War on Terrorism": 12-month Audit and Future Strategy Options*, Oxford Research Group, September 2002.

4 W. L. Eubank and L. Weinberg, 'Does Democracy Encourage Terrorism?', *Terrorism and Political Violence*, Vol. 6, Number 4, winter 1994.

5 P. Chalk, "The Liberal Democratic Response to Terrorism," *Terrorism and Political Violence*, Vol. 7, Number 4, winter 1995.

6 P. Wilkinson, *Terrorism and the Liberal State* (Macmillan, London, 1986).

7 P. Chalk, "The Liberal Democratic Response to Terrorism," *Terrorism and Political Violence*, Vol. 7, Number 4, winter 1995.

8 P. Wilkinson, *Terrorism and the Liberal State*, (Macmillan, London, 1986).

9 P. Chalk, "The Liberal Democratic Response to Terrorism," *Terrorism and Political Violence*, Vol. 7, Number 4, winter 1995.

10 J. Rotblat, *Nuclear Radiation in Warfare* (Taylor and Francis, London, 1981).

11 www.guardian.co.uk/g2/story/0,3604,769634,00.html www.theenolagay.com

12 G. Thomas and M. M. Witts, *Ruin from the Air: The Atomic Mission to Hiroshima* (Hamish Hamilton, London, 1977).

13 K. Osamu, *Children of Hiroshima*, Publishing Committee of "Children of Hiroshima," Tokyo, Japan, 1980.

14 W. Barnaby, *The Plague Makers* (Vision Paperbacks, London, 1999).

15 F. P. Horn and R. G. Breeze quoted in "Agriculture and Food Security," T. W. Frazier and D. C. Richardson (eds.), *Food and Agricultural Security*, Annals of the New York Academy of Sciences, Vol. 894, 1999.

16 D. Isenberg, *Quick Reference Guide to Biological Technology Equipment* (British American Security Information Council (BASIC)).
www.basicint.org/nuclear/biological/ORG2BW.htm

17 J. Simpson, *From the House of War: John Simpson in the Gulf* (Hutchinson, London, 1991).

18 *The Strategic Defence Review, A New Chapter* (Ministry of Defence, London, 2002).

19 Private communication between Hannes Alven and Frank Barnaby, 1979.

20 R. F. Mould, *Chernobyl Record: The Definitive History of the Chernobyl Catastrophe* (Institute of Physics Publishing, Bristol, 2000).

21 "Pyongyang goes for broke," Jane's Intelligence Review, March 2003.

22 Quotes are from the Treaty (BWC).

23 *Measures for controlling the threat from biological weapons* (The Royal Society, London, 2000).

24 Quotes are from the Treaties.

25 *ibid.*

26 George Shultz, US Secretary of State, speech on January 7, 1989 at the International Chemical Disarmament Conference, Paris.

27 Tony Blair, Prime Minister, speech on January 8, 2003 made to British ambassadors in London (Queen Elizabeth Centre).

28 J. Despres, "Intelligence and the Prevention of Nuclear Terrorism," in P. Leventhal and Y. Alexander (eds.), *Preventing Nuclear Terrorism* (Lexington Books, Mass., 1987).

29 The uranium threat, http://www.nci.org/new/about-nci.htm

30 J. C. Mark, T. Taylor, E. Eyster, W. Maraman and J. Wechesler, "Can Terrorists Build Nuclear Weapons?" in P. Leventhal and Y. Alexander (eds.), *Preventing Nuclear Terrorism* (Lexington Books, Mass., 1987).

31 A. B. Lovins, *Nuclear Weapons and Power-Reactor Plutonium*, Nature, London 283, 817-823 and typographical corrections, 284, 190, 1990.

32 Excerpts of Yigal Amir sentencing decision, March 27, 1996. www.israel-info.gov.il/mfa/go.asp?MFAHOIp30

33 P. Wilkinson, "Extreme Right-Wing Violence: Violence and Terror and the Extreme Right," *Terrorism and Political Violence*, Vol. 7, Number 4, winter 1995.

34 B. Hoffman, "Terrorist Targeting: Tactics, Trends and Potentialities," *Terrorism and Political Violence*, Vol. 5, Number 5, summer 1993.

35 E. Karmon, "The Role of Intelligence in Counter-Terrorism," paper presented at a conference on *Intelligence in the 21st Century*, at Castle of San Marino, Priverno, Italy, February 14–16, 2001.

36 J. Kaplan, "Leaderless Resistance," *Terrorism and Political Violence*, Vol. 9, Number 3, autumn 1997.

37 US Department of State, *Countering The Changing Threat of International Terrorism*, Report of the National Commission on Terrorism, US Congress, June 5, 2000. www.news.findlaw.com/cnn/docs/crs/natcomterr20601.pdf

38 I. O. Lesser, "Countering the New Terrorism: Implications for Strategy," in I. O. Lesser, B. Hoffman, J. Arquilla, D. F. Ronfeldt, M. Zanini, B. M. Jenkins, *Countering the New Terrorism*, RAND, California, 1999.

39 M. Bunn, *The Next Wave: Urgently Needed New Steps to Control Warheads and Fissile Material*, A joint publication of Harvard University's Project on Managing the Atom and the Non-proliferation Project of the Carnegie Endowment for International Peace, Cambridge, MA., and Washington, D.C., April 2000.

40 *ibid.*

41 Europol Convention on www.europa.eu.int/scadplus/leg/en/lvb/114005.htm

42 E. Marotta, "Europol's Role in Anti-Terrorism Policing," *Terrorism and Political Violence*, Vol. 11, Number 4, winter 1999.

43 E. Karmon, "The Role of Intelligence in Counter-Terrorism," paper presented at a conference on *Intelligence in the 21st Century*, at Castle of San Marino, Priverno, Italy, February 14–16, 2001.

44 M. El Baradei, speech at Conference on Radiological Dispersion Devices, International Atomic Energy Agency, Vienna, 2002.

45 M. Chirico, *Changing Preconceptions of the Nuclear Terrorism Threat:*

A Case Study of the Aum Shinrikyo Cult, Program in Science, Technology, and International Affairs, Georgetown University, USA.
www.georgetown.edu/sfs/programs/stia/students/chiricom.htm

46 W. Barnaby, *The Plague Makers* (Vision Paperbacks, London, 1999).

47 K. Alibek, *Biohazard* (Hutchinson, London, 1999).

48 T. Bartfai, S. J. Lundin and B. Rybeck, "Benefits and threats of developments in biotechnology and genetic engineering," in *World Armaments and Disarmament* (SIPRI Yearbook 1993, Oxford University Press, 1993).

49 V. Nathanson, speech at World Medical Congress, South Africa, 1996.

50 *"Genetic weapons: Could the latest research add a terrifying new dimension to warfare?,"* Foreign Report, Jane's Information Group, March 14, 1996.

51 M. M. Pollitt. *"A Cyberterrorism Fact or Fancy?,"* Proceedings of the 20th National Information Systems Security Conference, 1997, pp. 285-289.
www.crime-research.org/eng/library/Cyber-terrorism.htm

52 *Computers at Risk: Safe Computing in the Information Age* (National Academy of Science Press, Washington, DC, 1991).

53 *Countering the Changing Threat of International Terrorism*, Report of the National Commission on Terrorism, US Congress, Washington, DC, 2000.

54 D. E. Denning, *Activism, Hacktivism, and Cyberterrorism: The Internet as a Tool for Influencing Foreign Policy*, February 4, 2000.
www.nautilus.org/info-policy/workshop/papers/denning.html
D. E. Denning, *Cyberterrorism*. Testimony before the Special Oversight Panel on Terrorism Committee on Armed Services US House of Representatives, May 23, 2000.

Index